DO YOU KNOW—

—What handsome leafy plant will repel insect pests?

—Why you should give your plants a drink of tea?

—How much artificial light will perk up listless leaves?

—What common garden pest is struck dead by . . . a potato?

—What can cause yellow or whitish rings on leaves?

—What shocking remedy will make the Cape Primrose bloom?

This quick and easy rescue handbook for your sick houseplants offers complete care for 200 popular houseplants . . . identification guides for plants and pests . . . low-cost plant products—and home remedies that really work. Drawing on his years of experience as a professional nurseryman and plant writer, Jack Kramer shares his trade secrets to help you keep all your plants healthy and happy. His seasoned advice is a must for beginners and expert growers alike.

FIRST AID FOR PLANTS

JACK KRAMER, former owner of the Garden District plant shop in Napa Valley, California, is an experienced nurseryman and veteran plant writer. His books include *Easy Plants for Difficult Places*, *The Everest House Complete Book of Gardening*, *Patio Gardens*, and *Plant Language*.

First Aid
for
Plants

Jack Kramer

A PLUME BOOK

NEW AMERICAN LIBRARY

A DIVISION OF PENGUIN BOOKS USA INC., NEW YORK

PUBLISHED IN CANADA BY
PENGUIN BOOKS CANADA LIMITED, MARKHAM, ONTARIO

NAL BOOKS ARE AVAILABLE AT QUANTITY DISCOUNTS WHEN
USED TO PROMOTE PRODUCTS OR SERVICES. FOR INFORMATION
PLEASE WRITE TO PREMIUM MARKETING DIVISION,
NEW AMERICAN LIBRARY, 1633 BROADWAY, NEW YORK.
NEW YORK 10019.

PLUME TRADEMARK REG. U.S. PAT. OFF. AND FOREIGN COUNTRIES
REGISTERED TRADEMARK—MARCA REGISTRADA
HECHO EN DRESDEN, TN.

SIGNET, SIGNET CLASSIC, MENTOR, ONYX, PLUME, MERIDIAN and
NAL BOOKS are published in the United States by
New American Library, a division of Penguin Books USA Inc.,
1633 Broadway, New York, New York 10019, in Canada
by Penguin Books Canada Limited,
2801 John Street, Markham, Ontario L3R 1B4

Library of Congress Cataloging-in-Publication Data

Kramer, Jack, 1927–
    First aid for plants / Jack Kramer.
        p.   cm.
    ISBN 0-452-26261-5
    1. House plants—Diseases and pests—Control.   I. Title.
SB608.H84K69   1989
635.9'65—dc19   88-28910
                                                     CIP

First Printing, April, 1989

1   2   3   4   5   6   7   8   9

PRINTED IN THE UNITED STATES OF AMERICA

ACKNOWLEDGMENT

My thanks to the many people who visited my plant shop, the Garden District, in the Napa Valley, California, from 1982 to 1987. It was their many, many questions—most all about plant health—that prompted this book. And of course thanks to Hugh Rawson and my editor, Alexia Dorszynski, who nurtured the original idea to fruition.

Contents

Introduction: No More Sick Plants!   ix
How to Use This Book   xi

PART 1

The Plant World

Plant Groups   5

  Aroids   7
  Begonias   19
  Bromeliads   29
  Bulbs   41
  Cacti and Succulents   49
  Ferns   63
  Geraniums   75
  Gesneriads   89
  Orchids   105
  Palms   121
  Other Plants   133

PART 2

The Emergency Plant Protection Guide

How Plants Grow   147

Dealing With Plant Troubles   149

Afterword   167

A Word on Terms   168

Leaf Identification Guide   170

Pest Identification Guide   171

Introduction:
No More Sick Plants!

First Aid for Plants tells you how to save your plant's life—whether it is ailing or seriously ill. If you have a philodendron with limp leaves, an orchid with metabolism trouble, or a palm that looks terminal, treatment to help your ailing plant is here—in clear, nontechnical terms. I have included remedies, therapies, and some unorthodox techniques to put the green—or the blush of color—back into your plants.

Since I strongly believe in the dictum "know your patient," I've given a thorough explanation of the physiology of a plant so that you can better understand how the arteries, digestive system, and other internal functioning of plants work. Some plant rescue procedures include resuscitation chambers, ice-cube treatment, leaf amputation, and bypass root surgery.

With *First Aid for Plants* in hand there is no reason why your plants should shrivel, moan, groan, or swoon in despair. A healthy plant is a happy one.

How to Use This Book

Part one lists nearly two hundred of the most commonly grown plants in homes, based on reports by customers at my plant shop in five years. Also included are the health questions most asked about these plants. My remedies and cures for ailing plants are at the end of each plant group; details about health preparations and solutions and how to use them are to be found in part two. I do not guarantee a 100 percent cure factor; I myself lose a few plants now and then. So will you.

Gardeners or not, most people do recognize the difference between a fern and a philodendron, a cactus and a palm. Plants in this book are grouped by family, rather than listed alphabetically, so you won't have to thumb from aeschynanthus to streptocarpus to find the plant you are looking for. Where appropriate, I've used common names such as African-violet, rather than botanical names, such as *Saintpaulia*. Growers sell and people buy a philodendron, not an Aroid, a fern rather than a Polypodiaceae, a palm, not a Palmaceae. And most mail-order catalogs list plants by common names.

I've also provided leaf-pattern identification drawings to assist you in locating your ailing plant; they can be found in the appendix, at the back of the book.

PART 1

The Plant World

Our favorite houseplants come from all over the world: philodendrons from the jungles, orchids from the rain forests, some cacti and succulents from the blazing-hot deserts, and ferns from gladed shady areas. There are more than 400,000 species of flowering plants alone, 9,000 different kinds of ferns, and over 150,000 different orchids (hybrids included). Following is a brief listing of plants according to their native habitat:

West Indies: orchids, ferns, philodendrons, dieffenbachias, gesneriads

Central America: cycads, orchids, bromeliads, palms

Mexico: cacti, succulents, bromeliads, orchids

South America-Northern Area: palms, aroids, bromeliads

Southern Area: orchids, bromeliads, cacti, aroids, dieffenbachias, calatheas, begonias

Mountainous Areas: fuchsias, lapagerias, cacti, anthuriums, dieffenbachias, orchids

Mediterranean: oleanders, selaginellas, sempervivums

Equatorial Africa: dracaenas, begonias, clerodendrum, ficus, some palms

South Africa: streptocarpuses, ferns, host of flowering bulbs
such as haemanthus and zantedeschia

Malaysia: orchids, Ficus

Sumatra, Borneo, Java, Burma: gesneriads, begonias, some
orchids, ferns

Japan: palms, rhododendrons, lilies, orchids, ferns, camellias,
gardenias

China: azaleas, rhododendrons

Despite this immense variation in plant distribution, most of our best houseplants originated in the temperate zones, not in the tropics, as many people think.

At this point it is important to stress that *a definitive knowledge of plants' native habitats is NOT a prerequesite to growing those plants well indoors*. A very general knowledge of the geographical area and its climate is sufficient. Today's houseplants have been so hybridized and so cultivated, and there are so many new varieties of old species, that often the cultural requirements of the parent plant no longer apply. Hybridists are constantly breeding plants that adjust to adverse conditions; there is now even an orchid that blooms outdoors in 40° F temperatures. Again, a knowledge of climatic conditions for various houseplants in their native habitat is *not* really necessary.

The plant groups we discuss in this book are generally from the temperate to subtropical areas. The groups include aroids, begonias, orchids, and others. Within these groups you will find hundreds of plants to grow and cultivate healthily.

Plant Groups

Plant groups, which scientifically are called genus (singular) or genera (plural), are not complex; the way plants are named is relatively simple. Think of your own family and its members—sisters, brothers, uncles, aunts, and so on. You and your family and all the descendants share certain characteristics, as do the plants in a specific family. For example, consider the common house plant Bromeliad. The family is Bromeliaceae, the genus *Bromeliad*, and the species is the specific name. Look at *Bromelia balanse*. This plant is from the family Bromeliaceae; the genus is *Bromelia* and the species is *balanse*. This is the scientific name of the plant used worldwide.

In addition to species there are hybrids and varieties, which are basically improved forms of a species created by mating two different plants of the same family to produce a plant that excels in certain characteristics, such as large flowers or robustness or color or form. A good example is *Bromelia balanse*, 'Red Volcano.'

Most plants' names, whether in the families Bromeliads, Orchids, Gesneriads, and so on, are in Greek or Latin or are named after the person who discovered them, after the plant's characteristics, or the geographical area the plant came from. For instance, the *Tillandsia* and *Billbergia* groups of the Bromeliaceae family honor two Swedish botanists: Tillands and Billberg. A species name such as *circinnata* means twisted or rolled, referring to the plant leaves; *venezueleana* is a species that was discovered in Venezuela.

There are other groupings, such as order, class, division, and subdivision, within the plant world, but such technical discussion is not necessary for our purposes.

Following are the group names and descriptions of the most common plants grown by gardeners.

Aroids *(foliage plants)*

LIGHT: Partial
WATERING: Variable
FEEDING: Moderate

Aroids, a large group of plants, are considered foliage plants and used extensively in homes. The aroids have many good qualities: They need only low light and even moisture, and they are tough. This group includes both small and large plants, and leaf shapes vary considerably, from large, as in *Dieffenbachia*, to small, as in *Philodendron oxycardum*. Flowers are usually insignificant, except for the spathiphyllums, which have tall white spathes, (a large, leaflike part enclosing a flower cluster), and the anthuriums, whose spathes are red. Some of the more familiar aroids are Aglaonema (Chinese evergreen), Anthurium (flamingo flower), Dieffenbachia (dumb cane), Philodendron, and Spathiphyllum.

Aglaonema

I have been fooled so often by the performance of these that I do not know why I keep writing about them. Generally, I have had miserable luck with my aglaonemas, most expiring after a few years. Still, since some people admire them, here are the popular ones:

A. *commutatum pseudo-bracteatum*—bright green leaves that are splashed with yellow.

A. *pictum*—velvety leaves are spotted silver.

A. *simplex*—grows like a weed, even in water; it has dark green leaves.

GENERAL HEALTH CONDITIONS

Be sure to give these plants warmth (65°F) at night or growth will be a problem. Stems grow closely together and create ideal conditions for spider mites; be sure there is sufficient air ventilation.

AROIDS

Aglaonema commutatum
pseudo-bracteatum

Aglaonema pictum

Aglaonema simplex

Anthurium

This overlooked group of aroids makes an exquisite indoor decoration; the leaf patterns of some are quite dramatic. Although anthuriums need more warmth than many aroids (60°F at night), they are worth special consideration. Note that these plants like shade and do not like to be overfed.

A. andreanum—dark green leaves; certain varieties have white, red, orange, or pink spathelike flowers.

A. crystallinum—the velvety green leaves are beautifully veined in silver, but the flowers are insignificant.

A. scandens—climbing, with dark green foliage.

A. warocqueanum—long velvety leaves with pale green veins; an outstanding plant but tough to locate.

GENERAL HEALTH CONDITIONS

If you grow anthuriums with warmth and even moisture at the roots, pests should be no problem. Do be concerned about rot at the base of the plant, which is usually the by-product of too much watering. Include a good deal of sand in the soil mix.

AROIDS

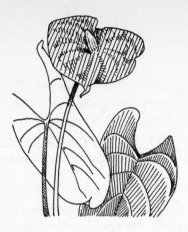

Anthurium andreanum

Anthurium crystallinum

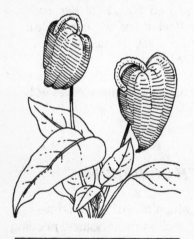

Anthurium scandens

Anthurium warocqueanum

Dieffenbachia

This is another fine group of aroids. The plants are graceful with large ornamental leaves marked with yellow or white. Dieffenbachia prefer light shade; allow soil to dry out between waterings. Don't panic when bottom leaves occasionally die back on mature specimens; this happens naturally. Remove and discard.

D. amoena—very large oval shiny green leaves, marked with white feathering along the veins. *D. amoena* 'Exotica' has dark green leaves dramatically marked with white splotches.

D. picta—there are some very good hybrids of this oval-leaved plant. *D. picta* 'Rudolph Roehrs' has bright green leaf margins, with the rest of the leaf a pretty greenish yellow. *D. picta* 'Superba' has creamy white leaves.

D. splendens—this dumb cane has velvety green leaves with white dots.

GENERAL HEALTH CONDITIONS

One of the more amenable groups of aroids, dieffenbachias are almost insect-free and grow well in most homes, needing few special considerations. They are a good choice for beginning gardeners.

AROIDS

Dieffenbachia amoena

Dieffenbachia picta

Dieffenbachia splendens

Philodendron

Some philodendrons have large leaves, others small ones. Many are vining plants, and a few are self-heading (ending with a dense growth of spiky flowers) or have a rosette growth. Philodendrons have long been popular in spite of the fact that they are not easy to grow because they require good air movement, adequate humidity, and careful watering. Here are some less demanding varieties:

P. andreanum—fast-growing vinelike plant with 10-inch-long dark green leaves.

P. bipinnatifidum—a somewhat different philodendron, with very pretty deeply scalloped leaves.

P. hastatum—perhaps the most popular philodendron, this 10-inch-tall plant has arrow-shaped leaves.

P. wendlandii—this plant, with its self-heading rosette growth and large and broad dark green leaves, resembles the bird's-nest fern.

GENERAL HEALTH CONDITIONS

Grow the plants in a somewhat moist and shady location, but avoid keeping the soil too soggy or too dry. Provide good air circulation to prevent red spider attack. Wash down leaves with a damp cloth every month or so.

AROIDS

Philodendron andreanum

Philodendron bipinnaitifidum

Philodendron hastatum

Philodendron wendlandii

Spathiphyllum

These aroids have been around a long time. Improved varieties such as Spathiphyllum floribundum 'Mauna Loa', with its large white spathes, have beefed up the demand. Generally the evergreens are easy to grow if in a very bright location and kept on the dry side. Recent offerings include the following:

S. floribundum—plain green leaves and white flowers.

S. floribundum 'Marion Wagner'—different because of its quilted leaves.

S. floribundum 'Mauna Loa'—satiny dark green leaves contrasted with large white spathelike flowers.

GENERAL HEALTH CONDITIONS

Mealybugs are a constant threat to these old-time favorites, so be on the alert and use precautions as cited in part two. Occasionally rot at the baseline develops, so do not overwater.

AROIDS

Spathiphyllum floribundum

Spathiphyllum f. 'Marion Wagner'

Spathiphyllum f. 'Mauna Loa'

Begonias

LIGHT: Partial sun
WATERING: Keep soil evenly moist
FEEDING: Moderate

These old-fashioned favorites are still popular today because the plant family is tremendously versatile. Begoniaceae leaves come in a variety of shapes, from star-shaped to button-shaped, and there are many different growing habits as well as several new flowering forms.

Most begonias (other than the true miniatures) are medium-sized plants, from 20 to 30 inches tall. Leaves may be large, as in the angel wings, or small, as in the miniature *B. boweri*. Most varieties have a colorful display of flowers in the summer and fall, from soft pink to orange to white. All the following types of begonias are perfect for growing: angel-wing and other fibrous begonias, basket begonias, hairy-leaved (hirsute) and rhizomatous begonias, and miniature begonias.

Angel-Wing and Other Fibrous Begonias

These easy-to-grow plants have leaves that are shaped like angel wings and that alternate on the stem. In the summer the plants carry bowers of very pretty flowers that last about a month.

In temperate climates, the angel-wings can grow outdoors in the sun. Soak the plants and then let them dry out.

B. coccinea—these large plants—up to 60 inches high—are worth the space they occupy; the canelike stems carry coral-red flowers almost all summer long.

B. 'Coraline de Lucerne'—the favorite spotted angel-wing begonia; the pale green leaves have white spots and pendant clusters of pink flowers.

B. 'Grey Feather'—grows as fast as a weed and has dark green, almost gray, leaves; the flowers are white with a pink hue.

B. 'Rieger'—a fine hybrid with orange or red flowers that last many weeks.

GENERAL HEALTH CONDITIONS

Angel-wings are relatively insect-free; their branching growth allows good air circulation, and the plants have an inborn tolerance to pests. Occasionally mealybugs may attack, but they are easy to eradicate (see part two). Plants tend to become leggy after a few years and then need to be cut back to about 8 to 12 inches. Avoid feeding angel-wings too much or too often—moderate feeding (twice a month with a 20-10-10 food) keeps plants healthy.

BEGONIAS

Begonia coccinea

Begonia 'Coraline de Lucerne'

Begonia 'Grey Feather'

Begonia (Reiger hybrid)

Basket Begonias

The tuberous (with bulblike roots) begonias are usually considered *the* basket begonias, but we will look at other types of begonias, with different leaf structures and flower forms. These begonias have fewer flowers than the tuberous kinds, and their variety of leaf patterns and shapes makes them unique decorations. Certain of these begonias can be grown outdoors if they are in protected areas where temperatures do not fall below 45°F. Indeed, the basket begonias make stunning outdoor statements.

B. 'D'Artagnan'—a trailing begonia, sometimes called the queen of the basket begonias; the forest-green, somewhat hairy foliage and bowers of pink or whitish flowers are quite showy.

B. 'Elsie M. Frey'—I have grown this basket begonia for 20 years. The plant is easy to care for, and it has metallic green-red-lined leaves and lovely bowers of pink flowers.

B. 'limminghei'—this winter-blooming begonia has shiny green leaves and coral-colored flowers. The plant likes warmth.

B. 'President Carnot'—popular because it blooms readily; the leaves are copper green, and the flowers can vary as to color, although they are usually pink.

GENERAL HEALTH CONDITIONS

If you grow the basket-type begonias in a slightly acid soil, with excellent drainage and plenty of air circulation, they should be healthy, and healthy plants are rarely bothered by insects. If there is an occasional attack of red spiders, fight the insects with nicotine "soup" (see part 2, page 141) or alcohol. Another good "cure" often prescribed by gardeners is to rub the affected leaves with powdered charcoal.

BEGONIAS

Begonia 'D'Artagnan'

Begonia 'Elsie M. Frey'

Begonia 'limminghei'

Begonia 'President Carnot'

Hairy-Leaved (Hirsute) and Rhizomatous Begonias

These big, furry-leaved begonias have been favorites for decades because of their wide range of leaf shape and texture. Many varieties have large masses of small flowers. Feed the plants regularly, more in the summer than in the winter.

B. erythrophylla—the beefsteak begonia; the round leaves are green on top, red underneath, and the flowers are a pretty pink.

B. 'Loma Alta'—highly attractive, with large, scalloped, plush red leaves.

B. 'Maphil'—sometimes called the Cleopatra begonia. The leaves are splashed with gold, chocolate brown, and chartreuse, and the plant bears pink flowers in the spring. The Cleopatra begonia makes a good basket plant.

B. metallica—hairy, dark green, purple-veined leaves; a good garden plant in the summer.

GENERAL HEALTH CONDITIONS

Because of the hairy leaves, insect infestation, such as mealybug and spider mite that is not caught quickly can result in a very messy situation. Many times it is too late to remedy the condition because any washing or other precautions mentioned in part two do not work. Be alert.

BEGONIAS

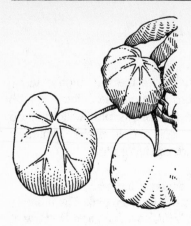

Begonia erythrophylla

Begonia 'Loma Alta'

Begonia 'Maphil'

Begonia metallica

Miniature Begonias

These begonias are small—no more than 10 inches tall—pretty, and easy to grow. I started my begonia collection with the miniatures years ago. Give the plants partial light, feed them moderately, and keep the soil evenly moist. Miniature begonias always provide fine indoor decoration.

B. boweri—the famous eyelash begonia; the green leaves are stitched black at the edges, and the flowers are pink. The many varieties include *B.* 'Bow Arriola', *B.* 'Bow Joe', and *B.* 'Bow Nigra'.

B. 'Chantilly Lace'—the beautiful pale green cupped leaves have delicate eyelash hairs at their edges.

B. foliosa—one of my favorites, this begonia looks like a tiny weeping willow, with arching stems and diminutive green leaves.

B. 'Red Berry'—a miniature folliage begonia; the wine-red foliage is quite dramatic.

GENERAL HEALTH CONDITIONS

As a rule, these liliputian plants are fairly resistant to pests and disease. They may occasionally be attacked by aphids or scale but can easily be controlled with measures mentioned in part two.

BEGONIAS

Begonia boweri

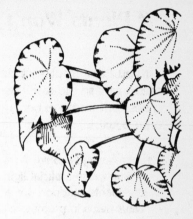

Begonia 'Chantilly Lace'

Begonia foliosa

Begonia 'Red Berry' (rex)

If Plants Won't Bloom:

If you have trouble getting begonias to bloom for you, try some of these old-fashioned remedies. I can make no guarantees, but in my experience, treated plants should produce some flowers.

- *Angel-wing begonia:* These plants need sun; if they don't get enough, use artificial light (see part two). You can also put the plants outdoors for a summer vacation; this almost always helps to promote blooming.

- *Rhizomatous and hairy-leaved begonia:* These plants are beautiful even if they don't bloom, but flowers provide satisfaction to many gardeners. For flowers on these begonias, try putting them in the deep-freeze at night, which is to say grow them quite cool (about 45–50°F) for about three weeks.

- *Basket begonia:* This is a very diversified group, so making these plants bloom is difficult. In general, try heavy feeding, and if that does not work, soak the plants in a sink of water for about 10 to 15 minutes.

- *Miniature begonia:* The secret to success with the mighty mites is to start with very good plants rather than bargain-basement specials. The trick is to buy high-quality miniatures; they will bloom with ample sun.

Bromeliads

LIGHT: Sun
WATERING: Keep vase filled
FEEDING: Light

With their colorful bracts (modified leaves growing at the base of the plant) and decorative foliage, the bromeliads are fast becoming a favorite indoor plant. These air plants ("epi-phytes," which means they grow on other plants), native to South America, are sometimes called pineapple plants. Most bromeliads are mid-size, to about 30 inches, and their growth is either vase- or rosette-shaped. The leaves are broad and solid green, banded, spiny, or grassy in appearance. The plants bloom in the summer and the winter. Some varieties have large crowns of tiny flowers on tall stems; others have vividly colored bracts, from bright cerise to orange. Flush out the water in the center of the vase every third day and keep the plant in good bright light. This colorful plant family offers the following excellent houseplants: Aechmea (silver urn plant), Guzmania Neoregelia (volcano plant), Tillandsia (tuft plant), and Vriesea (flaming sword).

Aechmea

These handsome, vase-shaped plants have banded or plain green leaves. The tall stems are crowned with hun-dreds of tiny flowers in bright and garish colors; flowers bloom only on new growth. Many varieties have colorful berries after the bloom cycle.

A. chantinii—this queen of the bromeliads has arching, toothed leaves that are beautifully marked with green, white, and silver. The flower spike is erect, tall, and branching, with yellow-edged red bracts that last for months.

A. fasciata—the most popular bromeliad grown. The broad, gray-green leaves are handsomely marked with silver bands, and hundreds of tiny blue flowers to a crown are carried in bright pink bracts on a tall stem. The variety *A. fasciata* 'Silver King' has heavily marked leaves and larger crowns of flowers.

A. fulgens—this plant is desirable because the small blue flowers on a pendant scape (hanging, shell) are followed by red berries that last 3 to 4 months. The arching leaves with spiny edges are dusted with gray.

A. 'Maginali'—a somewhat large plant (to 40 inches) with purplish foliage marked with white. The large, pendant scapes of bright cerise bracts and small whitish flowers are followed by purple berries that last for many months.

GENERAL HEALTH CONDITIONS

Aechmeas are healthy plants and can tolerate a certain amount of neglect before they become prone to sickness. Bottom leaves may brown off from too much water at the soil. Occasionally mealybugs attack the less-succulent-leaved aechmeas. Infrequent watering might result in leaves somewhat limp to the touch, and insufficient light can turn foliage pale. None of these problems is difficult to remedy; indeed, it is very rare for an aechmea simply to shrivel up and die.

BROMELIADS

Aechmea chantinii

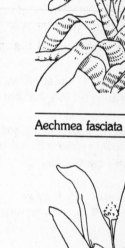

Aechmea fasciata

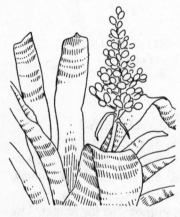

Aechmea fulgens

Aechmea 'Maginali'

Guzmania

With their fiery orange or red crowns of leaf bracts and rosette growth, the guzmanias are handsome plants for special accent. Keep the leaf cups filled with water and do not feed the plants. After they bloom, cut off the offsets (baby plants) when they are 4 inches tall and then pot each offset separately. In time, discard the mother plant.

> G. *lingulata major*—an attractive plant with large star-shaped crown of orange flowers that last for weeks.

> G. *monostachia*—a showy plant with satiny green leaves, a pokerlike flower spike with white flowers, and green bracts stenciled with maroon lines.

> G. 'Orangeade'—an improved variety of G. *lingulata*, with a very large, brilliant orange-red flower crown.

GENERAL HEALTH CONDITIONS

Watch the leaf cups—make sure they are always filled with water and that the water is flushed out several times a week. Occasionally, mealybugs may create havoc with the thin foliage. Aphids, too, sometimes put in their bid for dinner. Use precautions as listed in part two.

BROMELIADS

Guzmania lingulata major

Guzmania monostachia

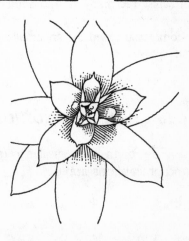

Guzmania 'Orangeade'

Neoregelia

Favorite indoor plants for generations in Europe, neoregelias are medium-sized (about 24 inches across) and grow in a flattened rosette shape. The foliage is plain green, striped, or banded or spotted. At bloom time the center of the plant turns fiery red. Keep the plants in good light and do not feed them. Keep the cup filled with water.

N. carolinae—the most popular neoregelia, with dark green leaves and a brilliant red center.

N. 'Painted Lady'—this plant has rosette of dark green leaves suffused with brownish red markings and a red center.

N. spectabilis—commonly called the fingernail plant; its spine has leaves tipped cerise.

GENERAL HEALTH CONDITIONS

Neoregelias have a growth habit similar to guzmanias, and the same rules and precautions apply.

BROMELIADS

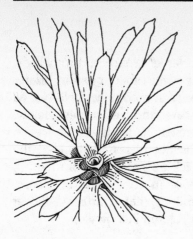

Neoregelia carolinae

Neoregelia 'Painted Lady'

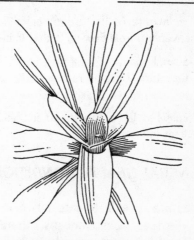

Neoregelia spectabilis

Tillandsia

The tillandsias form a very large group of bromeliads. Some plants have a palmlike growth; others grow in bizarre shapes, with twisted narrow leaves in a compact ball. Not all plants are handsome, but many are, and we list these below. Tillandsias need very good ventilation, fresh air, and even moisture all year to thrive. It is not necessary to feed the plants because most of them pick up moisture from the air. You can grow plants on slabs or in containers. In bloom, these plants (with a few exceptions) have thrusting flower stalks with small, very vibrantly colored little flowers.

T. argentea—a small plant with soft gray leaves in tufted growth and pink flowers.

T. cyanea—the favorite of the group; palmlike foliage and growth and heavenly purple flowers in the winter.

T. ionanthe—a small plant with a 2-inch ball of tufted leaves, and centers that turn brilliant red as they bloom in the spring.

T. juncea—somewhat large (up to 10 inches); yellow and red floral bracts in the summer.

GENERAL HEALTH CONDITIONS

The tillandsias are difficult to rate when it comes to pest problems and disease, mainly because there are so many different varieties of the plant—hundreds in fact. Since most tillandsias have leaves that are succulent in texture, insects will not harm them. However, I have found that on several plants scale has developed. Stay alert and use precautions as outlined in part two if you see any signs of the miniature "armored tanks."

BROMELIADS

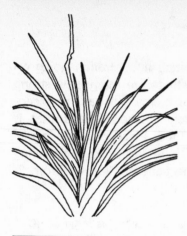

Tillandsia argentea

Tillandsia cyanea

Tillandsia ionanthe

Tillandsia juncea

Vreisea

Because there has been so much hybridization going on within this group of bromeliads, there are many new strong varieties, with tall flower bracts on stems. Vriesias are exceptionally beautiful, with exquisitely marked foliage that is banded, splotched, spotted, or lined. Place plants in partial light and do not feed them. There are several flowering types with orange bracts, but the patterned-leaved bromeliads are my choice.

V. bituminosa—this plant has a large rosette (as large as 40 inches), with broad blue-green leaves tipped purple.

V. carinata—this popular rosette grows to 24 inches, with orange bracts. Brilliant.

V. fenestralis—very handsome, with a 40-inch rosette of green leaves delicately marked darker green and lined with purple.

V. hieroglyphica—a 40-inch rosette of exquisite beauty, this plant has apple-green leaves banded darker gray-green as well as a tall flower spike.

GENERAL HEALTH CONDITIONS

The vriesias are considered exquisite foliage plants because of their striped and blotched foliage. People love the foliage color, and insects love the foliage. Watch out for aphids and scale mainly, and use measures outlined in part two to eradicate the trouble.

BROMELIADS

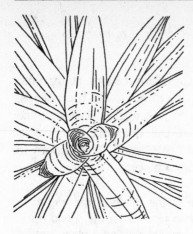

Vriesea bituminosa

Vriesea carinata

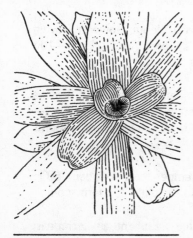

Vriesea fenestralis

Vriesea hieroglyphica

If Plants Won't Bloom:

Many people believe that bromeliads bloom every year from the same plant, but that is incorrect. Most bromeliads, from aechmeas and guzmanias to neoregelias, bloom once and then the mother plant dies—but not until it produces offsets at the base. When these are planted, they bloom in about 2 years. To get bromeliads to bloom:

- *Aechmea:* The old wives' tale of the apple in the bag to make aechmeas bloom has never worked for me. I end up with a rotten apple. What has worked is using a plant food mixture of 10-30-20. Applied once a week in April and May, this should do the trick.

- *Guzmania:* Quite tough to make bloom; however, strong light sometimes work. I use a standard reading lamp—100 watts—placed about 20 inches above the plants.

- *Neoregelia:* Start with almost-mature plants, about 2 years old; put them outdoors in summer in almost direct sun for about 3 weeks, then take them back into the house. Blooms should start in a few months.

- *Tillandsia:* Lower nighttime temperatures to about 60°F; use even lower temperatures to stimulate bud formation.

- Vreisea: These need the same treatment as aechmeas; feed heavily for a while.

Bulbs *(flowering and foliage)*

LIGHT: Partial
WATERING: Moderate
FEEDING: Very little

This overlooked group of plants includes some stunning flowers, such as the Pineapple flower (*Eucomis*) and the famous Amazon lily (*Eucharis*). Bulbs are not difficult to grow and require only potting, watering, and minimal feeding. Use standard potting soil, and generally feed plants mildly once a month with 20-20-10 formula. Most bulbs need a rest for about 3 months each year; leave them in their pots. In this group I have included the following lesser-known bulbs instead of the better-known ones, such as daffodils and freesias: Allium (flowering onion), Caladium, *Clivia* (Kaffir lily), *Colocasia* (elephant's ears), *Eucharis* (Amazon lily), *Eucomis* (pineapple lily), *Gloriosa* (Rothschild lily), *Haemanthus* (blood lily), *Vallota* (Scarborough lily), *Veltheimia*, and *Zantedeschia* (miniature calla lily).

Allium neapolitanum—the so-called "flowering onion" makes a fine pot plant, with green straplike leaves and umbels (clusters of flowers with stalks of equal length that spring from the same point) of starry flowers. Pot bulbs in sandy soil and give them plenty of light. Store the plants in a paper sack during the winter and then repot them in fresh soil in early March.

Caladium—Well known for its large, graceful foliage. There are dozens of varieties. Plant bulbs in a sandy soil, one to an 8-inch pot. Water the bulb sparingly at first, increasing the amount of moisture with growth. Let plants dry out in winter in a shady place.

Clivia miniata—an outdoor plant in temperate climates, this popular strap-leaved plant with orange flowers deserves to be grown indoors more than it is at present. Use large 10-inch pots, and let the soil dry out between waterings. Feed plants with a 10-10-5 solution every second week during the winter. Look for the Zimmerman hybrids and the famous Belgian hybrids with brilliant orange flowers. This dependable plant will bloom in April every year.

Colocasia—the "elephant's ear" has very handsome huge, scalloped dark green leaves that can reach 4 feet in length, so grow plants where there is plenty of space. Grow bulbs in large tubs, and water evenly. The plants prefer bright light.

Eucharis grandiflora—the "Amazon lily" has attractive straplike green foliage and bears clusters of pure white, highly fragrant flowers. Plants grow large (up to 4 feet tall), and flowers appear twice a year, in the spring and winter. Pot bulbs in a sandy soil, and cover them only halfway. Water bulbs sparingly until growth is well under way. In winter move the bulb (in the pot) to a cool place and water it sparingly. Repot bulbs in fresh soil every other year.

Eucomis punctata—handsome straplike foliage, with hundreds of tiny whitish green flowers blooming on a tall scape in the spring. Plant one bulb to a 5-inch pot in sandy soil, leaving the top part of the bulb exposed. You can grow the bulb in the same pot for many years, watering scantily in winter.

E. undulata—similar to *E. punctata*, but not as robust and with somewhat inconspicuous flowers; water scantily in winter.

Gloriosa rothschildiana—a Victorian favorite, this bulb produces fantastic twisted-petaled flowers with red lines. Grow these vining plants from a tuber: Start one to a pot in the spring. Give the tubers plenty of water, sun, and food (use a 20-20-10 formula every second week). In the winter let the tubers rest in a cool, dry place and then start them in 6 to 9 weeks in fresh soil.

Haemanthus coccineus—this "blood lily" is an excellent and unique plant: The flowers resemble a sphere of color. Start

BULBS

Allium neopolitanum

Caladium

Clivia miniata

Colocasia

the bulbs in sandy soil, leaving the tops of the bulbs exposed. Water bulbs sparingly until growth starts, and protect plants from the sun. *H. coccineus* grows all year, but is somewhat dormant in the summer.

H. Katherinae—the most popular blood lily, with 6-inch balls of blood-red flowers. Plant one bulb to an 8-inch pot in a fast-draining soil. Water bulb evenly and feed plants in spring when in active growth. Protect plants from sun. Reduce watering somewhat and grow almost dry in winter.

Vallota speciosa—this evergreen hails from South Africa; it has dazzling red flowers that appear before the foliage does. Plant one bulb to a 5-inch pot, with the nose of the bulb slightly above the soil. Keep the bulbs in a bright but not sunny place, and feed them every 2 weeks with a 10-10-5 solution from the spring through the summer, water scantily in winter. Vallotas are slow to come around, so they may not bloom until their second year.

Veltheimia—this plant from the Cape of Good Hope is one of the most dependable indoor bloomers you can grow: It almost always bears flowers at Christmastime. Yellowish red flowers appear at the top of erect stems. In late fall start one bulb to a pot in loose soil so that drainage is almost perfect. Once growth starts, heavy watering is needed. Feed bulbs with a 10-10-5 solution every 2 weeks, and grow plants in the sun. Let plants rest for 2 months after they flower, and then restart bulbs in fresh soil. *V. viridifolia* is the preferred plant in the family.

Zantedeschia—these often-overlooked miniature calla lilies are superb plants, blooming regularly in the summer. Start a few bulbs to a 5-inch pot. Water bulbs moderately, and feed them with a mild solution of 10-10-5 once a month. Keep plants in bright light. Good varieties are *Z. aethopica*, with white flowers, and *Z. rehmanni*, the best of the bunch, with lovely pink flowers.

Eucharis grandiflora

Eucomis punctata

Eucomis undulata

BULBS

Gloriosa rothschildiana

Haemanthus coccineus

Haemanthus Katherinae

Vallota speciosa

Veltheimia viridifolia

Zantedeschia aethopica

Zantedeschia rehmanni

GENERAL HEALTH CONDITIONS

As a rule, the majority of bulbous plants are exceedingly easy to grow, Most of them die down at some time of year, and if this resting period is observed, general houseplant care is all that is needed to keep plants trouble-free and insect-free. Occasionally, if you find mealybugs or aphids on grassy-type foliage, such as on *Eucomis* or *Eucharis*, sponge or wash off frequently and don't panic. Most of the bulb-type plants have great vigor and can fend for themselves when needed. Don't miss this group of plants—they offer great indoor beauty and are easy to care for and, as I've mentioned, are relatively insect- and disease-free.

If Plants Won't Bloom:

Again and again, we're told that bulbs have all the makings of the flowers within them. Maybe so, but a regular feeding program helps tremendously, too. Feed your bulbous flowering plants with a 10-20-20 mixture once growth has started. Use the feeding program for about 6 weeks at 10-day intervals for optimum results.

Cacti/Succulents

LIGHT: *Bright*
WATERING: *Variable*
FEEDING: *Moderate*

This immense group of plants includes specimens very good for decoration and others that are not worth your time. Although all cacti are succulents, not all succulents are cacti. The fleshy leaf is a fairly good indication of a succulent; usually, but not always, spiny leaves signify cacti. The group includes such favorites as the crown-of-thorns, poinsettia, Christmas cactus, and orchid cactus. Because many cacti come from the rain forest rather than the desert, plants need varied cultural conditions. The general rule with succulents and cacti is not to overwater them. If plants are healthy, they will rarely be bothered with insects or disease.

The various groups I'll describe here are the following: *Epiphyllum* (orchid cactus), *Euphorbia, Kalanchoe, Noctocactus* (ball cactus), *Rebutia* (crown cactus), and *Schlumbergera/Zygocactus* (Easter/Christmas cacti).

Epiphyllum

Long-time favorites of many gardeners, these rather ungainly plants bear mammoth and vibrantly colored flowers that stun the eye. These epiphytic plants need equal parts of fir bark and soil as their potting mix. Water plants heavily in the spring and summer but not as much the rest of the year. It is best to cultivate orchid cactus in hanging baskets or supported with stakes.

E. 'Conway Giant'—stunning 7-inch purplish red flowers.

E. 'Nocturne'—charming purple and white night-blooming flowers.

E. 'Parade'—a small but lovely plant with 5-inch pastel-pink flowers.

GENERAL HEALTH CONDITIONS

The leaves of *Epiphyllum* are tough and succulent in texture and not good eating, for most insects. A slight rest in winter with little water for about 6 weeks helps ward off premature spring growth.

CACTI/SUCCULENTS

Epiphyllum 'Nocturne'

Epiphyllum 'Parade'

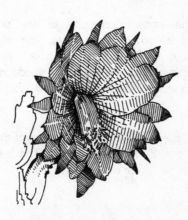

Epiphyllum 'Conway Giant'

Euphorbia

The euphorbias are not given as much attention as they deserve: Many make excellent houseplants. Euphorbias generally have somewhat succulent leaves and handsome red or orange flowers. The plants grow well, with minimal care. They are well worth space in your home, greenhouse, or protected garden.

E. keysi 'Pink Cloud'—attractive, with succulent leaves and coral-pink flowers in the winter.

E. milii—the most popular species; small dark green leaves and thick stems edged with spines.

E. pulcherrima—known commonly as the "poinsettia." The colorful bracts are red, pink, or white. The 'Mikkelsen' varieties are perhaps the best. Total darkness is not necessary to encourage bloom; place the plant in a shady location without any light for 6 weeks before plants bloom (they bloom around October).

E. splendens 'Bojeri'—the favorite "crown-of-thorns," a lovely dwarf that grows 20 inches in height. The leaves are spiny, and the plants have fine red flowers.

GENERAL HEALTH CONDITIONS

Generally Euphorbias are good robust plants—with the exception of poinsettias, which are occasionally attacked by mealybug. Follow preventatives in part two.

Euphorbia keysi 'Pink Cloud'

Euphorbia millii

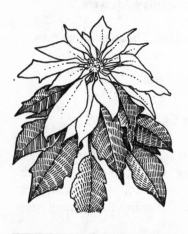

Euphorbia pulcherrima

Euphorbia splendens 'Bojeri'

Kalanchoe

These good but unusual plants are being seen more and more lately, and the original species, *K. blossfeldiana*, is now available in red, yellow, orange, or magenta. Kalanchoes are generally easy to grow: Give plants even moisture at the roots and feed them moderately, except in the winter, when they can be kept almost dry.

K. beharensis—unique, with fine, large, scalloped brownish green leaves that have a velvety texture.

K. blossfeldiana—the most popular of the group: small, succulent, dark green leaves and stems of colorful flowers.

K. tomentosa—called the "panda plant" because of its fuzzy and spotted gray-brown leaves.

GENERAL HEALTH CONDITIONS

Usually not bothered too much by insects, kalanchoes are sometimes attacked by mealybugs, which can easily be eradicated. Overwatering can cause bacterial rot at the stem base. Keep cats away from the plants; the milky sap attracts pets and can be harmful to them.

CACTI/SUCCULENTS

Kalanchoe beharansis

Kalanchoe blossfeldiana

Kalanchoe tomentosa

Notocactus

This is a fine group of cylindrical-shaped South American desert plants. Most plants are small—growing only to about 6 inches across—but all bear vibrantly colored flowers in the spring and summer. Best of all, you can count on the plants to bloom if you give them sufficient sun in the winter. Grow the plants in equal parts of soil and sand, and feed them with a 10-20-10 solution twice a month from February until July but not at all the rest of the year. Taper off the watering in midwinter: Apply moisture just once a month.

N. arachnites—a 6-inch globe, with heavily ridged and stunning yellowish green flowers.

N. leninghausii—a 4-inch globe, with somewhat spiny leaves that bear large yellow flowers.

N. rutilans—another small (up to 4 inches around) globe, this plant bears fine pink flowers.

GENERAL HEALTH CONDITIONS

As with most barrel-shaped cacti, these plants are relatively insect-free. If severely neglected, mealybugs may attack along the spine ridges. Handpicking with a toothpick is the only solution.

CACTI/SUCCULENTS

Notocactus arachnites

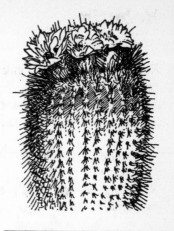

Notocactus leninghausii

Notocactus rutilans

Rebutia

Rebutia is the great little globe cactus, perfect for spot decoration because of its bright red flowers. The plants grow easily with minimal water and very mild feeding. Give *Rebutia* much light, but keep the temperature cool (50°F).

R. kupperiana—a small 4-inch globe of red flowers.

R. violaciflora—the small flattened 6-inch globe bears delightful purple flowers.

GENERAL HEALTH CONDITIONS

These are usually trouble-free plants—good, robust indoor subjects with few pests.

CACTI/SUCCULENTS

Rebutia kupperiana

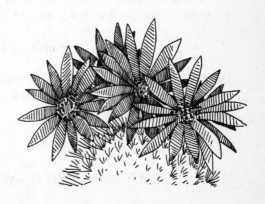

Rebutia violaciflora

Schlumbergera and Zygocactus

These plants have been interbred so much that it is almost impossible to detect which bloom in which seasons. For the sake of argument, *Zygocactus* is considered the Christmas cactus, *Schlumbergera* the Easter or Thanksgiving cactus. In any case, these are handsome plants that display a wealth of flowers in their season. Grow the plants evenly moist, and do not overfeed them—a mild solution all year-round is a wise choice.

S. bridgesii—a typical Thanksgiving cactus with red flowers; *S. bridgesii* 'Pink Perfection' has large pink, almost red, flowers.

Z. truncatus—called the "Christmas or crab cactus"; green segmented leaves and red flowers. *Z. truncatus* 'Gertrude Beahm' has bright red flowers, *Z. truncatus* 'Orange Glory' has pale orange flowers, and the flowers of *Z. truncatus* 'Symphony' are a delicate shade of orange.

GENERAL HEALTH CONDITIONS

I have rarely heard of these cacti (which are epiphytic jungle plants) being bothered by insects. Just remember they are rain-forest denizens—and keep them moist and shady.

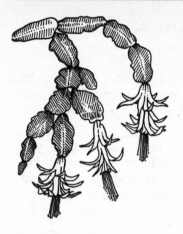

Schlumbergera bridgesii

Schlumbergera b. 'Pink Perfection'

Zygocactus truncatus
'Gertrude Beahm'

Zygocactus t. 'Orange Glory'

If Plants Won't Bloom:

Don't expect too many flowers from cacti and succulents grown indoors; however, there are a few unorthodox methods that might coax some varieties into bloom.

- *Epiphyllum:* If you want flowers from these beauties, you must give them a winter rest—and I do mean a rest. Put them in a shady place (leave them in their pots) in a cool location (50°F) for at least 8 weeks, and water only a few times.

- *Euphorbia:* Does the in-the-dark 6-week period for Christmas poinsettias really work? Sure, but it is not necessary to put them in a closet. Just make sure no artificial light strikes them at night. No real tricks here.

- *Kalanchoe:* With ample feeding kalanchoes bloom on schedule in early spring; if you want another crop of flowers, cut varieties such as *K. blossfeldiana* and its hybrids back to about 4 inches; when new growth starts, you begin another feeding program for a second crop.

- *Notocactus and Rebutia:* Light is the key factor for bloom in these plants. Supply additional artificial light. Even a 75-watt standard reading lamp, left on for about 5 hours at night, should promote bloom.

- *Schlumbergera/Zygocactus:* Follow the same in-the-dark periods as for the poinsettias.

Ferns

LIGHT: *Partial sun*
WATERING: *Heavy*
FEEDING: *Occasionally*

The fern was the workhorse of the Victorian indoor garden and the graceful note of the many outdoor gardens. Ferns are versatile foliage plants, handsomely lush and tropical looking, striking the perfect note as both indoor and outdoor decorations. Some ferns have languid fronds, whereas others have somewhat stiff leaves. Unfortunately, all this beauty demands attention: Ferns are prone to insect attack and will fail without the proper care. The plants do best in cool indoor situations (about 70°F during the day, 60°F at night) and need abundant water in warm weather. Many varieties, such as the Boston ferns, create drama when they are grown in baskets. Because many ferns are epiphytic rather than terrestrial, use equal parts of fir bark (the type sold for orchids) and soil as a potting medium. Do not feed ferns too often, as most react adversely to overfeeding. The following ferns will please avid gardeners:

Adiantum (maidenhair fern), *Asplenium* (bird's-nest fern), *Davallia* (rabit's-foot fern), *Lygodium* (climbing fern), *Nephrolepis* (Boston fern).

Adiantum

Most species of *Adiantum* are graceful and decorative, with delicate, lacy fronds and wiry black stems. These beautiful indoor plants are not overly large. Frequently *Adiantum* is used for indoor decoration where a lush verdant accent is needed. However, the plants do require some pampering. Protect these ferns from strong sun and excessive heat, and keep humidity at about 50 percent to thwart red spiders.

A. cuneatum (A. raddianum)—the most popular "maidenhair" fern, with lovely and delicate dark green fronds.

A. hispidulum—somewhat erect hairy fronds on stiff stems.

A. 'Sea Foam'—its lush growth makes this a very pretty fern.

A. tenerum—a large plant (up to 40 inches), with graceful, arching fronds that are sometimes tinged with pink. *A. tenerum* 'Wrightii' is a good, robust variety.

GENERAL HEALTH CONDITIONS

Adiantum is prone to mealybugs, which lurk in the fern's leaf axils. If you do not keep the soil evenly moist most of the year, the fronds will curl. And avoid bruising the tips of the fronds, as they will turn brown. *Adiantum* tolerates more feeding than most ferns. Keep the plants out of drafts, and avoid abrupt changes in temperature to avoid wilt.

FERNS

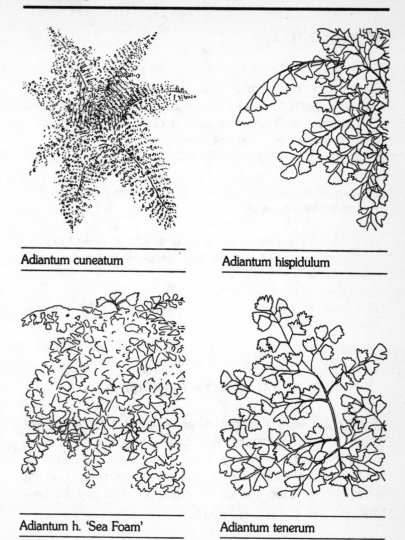

Adiantum cuneatum

Adiantum hispidulum

Adiantum h. 'Sea Foam'

Adiantum tenerum

Asplenium

These handsome ferns, which grow to 20 to 30 inches, have shiny green parchment-thin fronds. The large specimens are dramatic statements; the smaller plants, residing in, say, 6-inch pots, are generally not worthwhile. *Asplenium* is subject to attack by certain insects, such as scale and mealybug.

A. bulbiferum—wide fronds on black wiry stems, which tend to make it look ungainly.

A. nidus—the popular and very decorative "bird's-nest" fern; grows from a central core and has handsome and wide pale green fronds.

A. palmatum—an overlooked fine small fern, triangular fronds, papery texture.

GENERAL HEALTH CONDITIONS

Both these ferns are somewhat temperamental and need the best that nature can provide: dappled light, even temperatures, and careful watering. They must never be too dry or too wet, since rot will easily develop. Any brown spots on the leaves are not scale but seeds. Remove when dry for new plants.

FERNS

Asplenium bulbiferum

Asplenium nidus

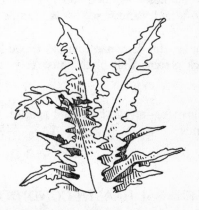

Asplenium palmatum

Davallia

Davallias are delightful plants, with fuzzy brown or gray rhizomes (rootlike stems) growing across the surface of the soil and very beautiful dark green lacelike fronds. These ferns thrive in open slatted wire or wooden baskets; they do not do well in standard containers. The plants can be grown outdoors in temperate climates. Some davallias grow very large, making them excellent accents.

D. bullata mariesii—a rather dainty plant, with brown rhizomes and lacy fronds. It is a good fern for outdoors because it can tolerate temperatures to 40°F.

D. fejeensis 'Plumosa'—a good variety of *D. fejeensis*; it has extremely dissected, narrow, segmented fronds.

D. pentaphylla—this fern has unusual broad-leaved fronds; the rootstock glistens with black hairs. The plant can grow large.

D. trichomanioides—an old-fashioned species, with lacy segmented fronds.

GENERAL HEALTH CONDITIONS

Davallia grows well in average home temperatures. These plants are not readily plagued by insects, although occasionally scale or mealybugs may attack, (see part two).

Davallia bullata mariesii

Davallia fajeensis 'Plumosa'

Davallia pentaphylla

Davallia trichomanioides

Lygodium

Lygodium is an unusual climbing fern that, like euphorbias, deserves more attention than it gets. With their lacy blue-green fronds, they survive in almost any situation—I have never had one succumb to ill health.

L. palmatum—a handsome fern with four to seven lobed fronds; needs support, such as a pole or a stick. It is sometimes called the Hartford fern.

L. scandens—a small-growing fern with 2-inch feathery blue-green fronds.

GENERAL HEALTH CONDITIONS

Lygodium is relatively free of problems; good care ensures a healthy plant that is rarely attacked by insects or disease. Grow plants on the cool side (60°F at night).

FERNS

Lygodium palmatum

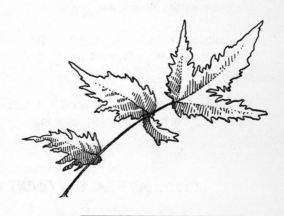

Lygodium scandens

Nephrolepis

This is the durable and popular of the fern family, but frankly, I do not think it deserves its reputation. The plants tend to shed when they are very mature and they're very particular when it comes to care: not too much water, but not too little, either. I consider the plants more trouble than they are worth, but people love them because of their lush look.

N. exaltata 'Bostoniensis'—the true Boston fern, with arching emerald-green fronds; it does best when grown in baskets.

N. exaltata 'Fluffy Ruffles'—very pretty, with its lacy and fluffy appearance, but only for about a year; then it seems to undergo a nervous breakdown.

N. exaltata 'Verona'—my choice in this group. The fern is rather robust, with somewhat small fronds (about 10 inches long).

N. exaltata 'Whitmannii'—a good newcomer because it is robust and has handsome 18-inch fronds.

GENERAL HEALTH CONDITIONS

As I've mentioned, these ferns are temperamental; they loathe drafts, fluctuating temperatures, complete dryness, and too much moisture. Occasionally scale will attack the plants, but brown spots on the backs of leaves are usually just spores (seeds).

Nephrolepis exalta 'Bostoniense'

Nephrolepis 'Fluffy Ruffles'

Nephrolepis 'Verona'

Nephrolepis 'Whitmannii'

Geraniums *(Pelargoniums)*

LIGHT: Partial sun
WATERING: Moderate
FEEDING: Moderate

Geraniums go in and out of favor, but through the years I have found that having a few of these choice plants indoors as pot plants or outside as bedding color is refreshing and worthwhile. The plants come in exquisite colors—pink, white, cerise, rose, red, and various shades and tints of these colors. As a group they are not difficult to grow, but they do have their requirements.

Most geraniums are medium-size plants (up to 40 inches), but outdoors some can grow into almost shrublike proportions measuring many feet. Indoors in pots, miniatures and semidwarfs are outstanding for color and grow well in cool temperature. (55°F at night), as do the ivy-leaved and fancy-leaved types. The most popular varieties are the following: fancy-leaved geraniums, ivy-leaved geraniums, Lady Washington geraniums, miniature and semidwarf geraniums, scented-leaved geraniums, and zonal geraniums.

Fancy-Leaved Geraniums

Fancy-leaved geraniums are just that—plants with splendid, multicolored leaves. Their colors run from bronze to almost black, silvery green, creamy white, brilliant pink, rust-red, yellow-green, and so on. There is an almost infinite variety, making this group a fascinating collector's choice. The plants originated in England in the mid-nineteenth century and need a somewhat shady place at the window where it is cool at night (55°F). Even watering is necessary, as is moderate feeding. Here are some suggestions:

G. 'Crystal Palace Gem'—yellow and bright green leaves; single red flowers (not pictured).

G. 'Filigree'—dwarf-type with tricolored silver leaves and pink flowers.

G. 'Happy Thought'—yellow in center of leaves, with green leaf margins and red flowers.

G. 'Jubilee'—red-brown coloring on yellow-green leaves; salmon-colored flowers.

G. 'Skies of Italy'—tricolored golden leaves, red flowers.

GENERAL HEALTH CONDITIONS

Foliage can be ruined by aphids or spider mites, and these do occasionally attack geraniums. Where temperature is high and humidity low, watch out for these pests and use the remedies mentioned in part two.

GERANIUMS

'Filigree'

'Jubilee'

'Happy Thought'

'Skies of Italy'

Ivy-Leaved Geraniums

The ivy-leaved geranium, technically known as *Pelargonium peltatum*, is so named because its leaves are peltate or like shields in shape. This group contains climbing, trailing, twisting, and twining varieties, which I find infinitely interesting and which can be used in many ways: in pots at windows, in hanging baskets, and on trellises. Flower colors vary with shades of white and red, and foliage may be variegated or plain green. Plants like a sunny place that is cool (55°F at night). Pinch plants back in late winter to promote many stems to grow.

P. 'Charles Turner'—double pink flowers; a fast climber.

P. 'Comtesse de Grey'—single light pink flowers.

P. 'Santa Paula'—double lavender-purple flowers.

P. 'Victorville'—double, dark-red flowers.

GENERAL HEALTH CONDITIONS

Bacterial leaf spot and stem rot sometimes hit these geraniums and are not easily cured. See part two for remedies for these problems.

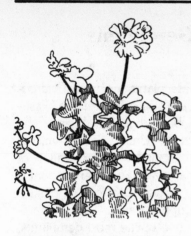

'Charles Turner'

'Comtesse de Grey'

'Santa Paula'

'Victorville'

Lady Washington Geraniums

Botanically speaking, all of these plants are from *Pelargonium domesticum* derivation and are also called Martha Washington types. The plants have been hybridized since the mid-nineteenth century—many in the United States—and at one time were also called regal geraniums. Hybridization of these plants goes so fast that what may have been current 5 years ago is obsolete today. The flowers are rose or pinkish marked with dark purple at the upper petals. These geraniums have stiff, rounded leaves, quite unlike most geraniums, and flowers that are petunialike in shape. The plants really need cool temperatures in winter to grow well (about 45°F at night), so an unheated room is the place for them. Good light is essential for blooming.

P. 'Dubonnet'—ruffled wine-red flowers.

P. 'Easter Greeting'—cerise flowers.

P. 'Holiday'—ruffled white flowers with crimson markings.

GENERAL HEALTH CONDITIONS

These geraniums are sometimes attacked by stem rot or black rot: Stems blacken and shrivel. Cut away such areas and hope for the best. Use sterile instruments when cutting, propagating, or handling plants, as the disease is spread by contaminated equipment. If white flies appear, use the remedies listed in part two to eliminate them.

GERANIUMS

Pelargonium 'DuBonnet'

Pelargonium 'Holiday'

Pelargonium 'Easter Greeting'

Miniature and
Semi-dwarf Geraniums

The small geraniums may be the most popular as indoor plants—they are charming and exciting. The height of these small plants varies with the gardener, but in my experience, miniatures grow to 3 inches, dwarfs to 6 inches, and semidwarfs to 8 inches. Plants grow slowly, and most varieties like a sunny spot; they are generally heavy feeders and require even watering—neither too dry nor too wet. Average home temperatures of 78°F by day and 10 degrees cooler at night are fine for healthy growth.

Miniature

P. 'Black Vesuvius'—dark leaves and orange-scarlet flowers (not pictured).

P. 'Fairy Tale'—white flowers with lavender centers (not pictured).

P. 'Imp'—dark leaves setting off single pink flowers (not pictured).

P. 'Saturn'—dark leaves and red flowers (not pictured).

Dwarf

P. 'Capella'—forest-green leaves and salmon-pink flowers (not pictured).

P. 'Lyric'—orchid-pink flowers (not pictured).

Semi-Dwarf

P. 'Dancer'—single salmon-colored flowers distinguish this somewhat large plant (not pictured).

P. 'Emma Hossler'—double white-centered, rose-pink flowers (not pictured).

GENERAL HEALTH CONDITIONS

These small cousins of larger geraniums suffer the same problems as their bigger relatives. Be on the alert and use appropriate remedies, as described in part two.

Scented-Leaved Geraniums

The scented-leaved geraniums have a wonderful fragrance, making them very desirable houseplants. The fragrance can be fruity or nutty, pungent or sweet. Foliage size varies from small—½-inch leaves—to 1-inch or more in size. Plants do well in a somewhat sunny location in average home temperatures with a 10-degree drop at night to approximately 60°F. Water judiciously and feed moderately.

Rose-Scented (Pelargonium graveolens)

'Dr. Livingston'
'Lady Plymouth'
'Little Gem'
'Snowflake'

Peppermint-Scented (Pelargonium tomentosum)

'Peppermint'
'Pungent Peppermint'

Lemon-Scented (Pelargoniun crispum)

'Gooseberry'
'Lady Mary'
'Orange'
'Prince Rupert'

Apple-Scented (Pelargonium odoratissimum) (not pictured).

'Apple'
'Old Spice'

GENERAL HEALTH CONDITIONS

These plants are only occasionally bothered by insects—mealybugs and whitefly—but these are easily controlled and the plants generally recover quickly.

GERANIUMS

Pelargonium Tomentosum

Pelargonium Crispum

Pelargonium Graveulens

Zonal Geraniums

These are the familiar old-fashioned type, with scalloped leaves and brilliantly colored flowers. They are derived from *Pelargonium hortorum* species. The plants adapt beautifully to indoor or outdoor growing. Flower colors include white, salmon, pink, red, lavender, and violet. There are single-flowered types, double-flowered, and even semidouble varieties. I have always grown a few pots of these zonals indoors in average home temperatures of 75°F by day, and about 60°F by night. I let plants dry out between waterings and keep them at an eastern or southern window.

P. 'Better Times'—single red flowers.

P. 'Dreams'—double salmon flowers (not pictured).

P. 'Fire'—single salmon flowers.

P. 'Harvest Moon'—single orange flowers.

P. 'Princess Fiat'—double light pink flowers (not pictured).

P. 'Snowball'—double white flowers (not pictured).

P. 'Starlight'—single white flowers (not pictured).

P. 'Summer Cloud'—double white flowers.

GENERAL HEALTH CONDITIONS

The favorites are perhaps so designated because they are not usually attacked by insects. Occasionally slugs or snails will feast on the leaves. Use appropriate snail-type baits as mentioned in part two.

'Better Times'

'Fire'

'Harvest Moon'

'Summer Cloud'

If Plants Won't Bloom:

Generally most geraniums bloom readily if they have some time outdoors. However, for indoor bloom there are a few notes that might help you to help the plants bear flowers:

- *Fancy-Leaved Geranium:* Usually these are grown for their beautiful foliage, but many have handsome flowers. To encourage blooming, give plants some supplemental (artificial) light, you need not get special lamps for this, however.

- *Ivy-Leaved Geranium:* The ivy-leaved varieties usually bloom quite well indoors without much coaxing if all conditions are good: watering, feeding, light, temperature. If you are having trouble getting flowers from these plants, increase feeding to four times a month using 10-20-20.

- *Lady Washington Geranium:* These plants have fine flowers, and you want them to bloom as much as possible. Of course, outdoors they will bloom their heads off, but if you have them indoors, be sure plants are cool at night—at temperatures to about 45° to 50°F—to encourage bud formation.

- *Miniature and Semidwarf Geranium:* As a rule these plants are not difficult to bring into bloom. If you are having problems getting them to bloom, drop the night temperature to about 50°F and be sure there is sufficient light.

- *Scented-Leaved Geranium:* Flowers are not generally important with the scented geraniums; they are grown for their wonderful fragrant leaves.

- *Zonal Geranium:* This large group of plants need plenty of water and feeding to bloom indoors, so feed, every other watering with 10-20-10 during growth. Provide ample light for them. Outdoors there is no special requirement, and plants usually bloom without much problem.

Gesneriads *(African violet, others)*

LIGHT: *Bright*
WATERING: *Keep soil evenly moist*
FEEDING: *Moderate*

The most popular gesneriad has long been the African violet (*Saintpaulia*), discovered by a German officer in 1892 in Tanganyika, East Africa. Today African violets and other gesneriads hardly resemble their original species because of the great strides in hybridization, which have produced stronger and better varieties. Most gesneriads are small- to medium-size plants, growing up to 24 inches in height, which makes them ideal for homes with limited space. The plants grow well in diffused light with adequate ventilation and watering. *Never* use very cold or ice water on the plants. Most gesneriads bloom in the summer and autumn, a few in the winter. Flowers may be tubular (up to 6 inches across), such as the gloxinia, or small (1 inch across) such as the African violet. There is a vast color palette: orange, red, pink, and purple. Most gesneriads have a common characteristic: hairy leaves in a rosette growth. This varied family offers: *Aeschynanthus* (lipstick vine), *Columnea, Episcia* (rainbow flower), *Hypocyrta* (goldfish plant), *Saintpaulia* (African violet), *sinningia* (gloxinia), and *Streptocarpus* (Cape primrose).

Aeschynanthus

Sometimes called *Trichosporum,* this lovely trailing epi-phyte has oval-shaped dark green leaves. Some varieties have small (1 inch) leaves. Plants produce a wealth of tubular flowers that are usually red but occasionally green or brown. It is best to grow these plants in hanging baskets.

A. *lobbianus*—a vine with fleshy oval leaves on trailing stems. Red bracts hold brilliant scarlet flowers that have a yellow throat. A. *lobbianus* 'Red Flame' has gorgeous red flowers.

A. *longiflorus*—perhaps the best of the "lipstick vines," it is a robust plant, with large, oval-shaped shiny dark green leaves and masses of large red flowers.

A. *speciosus*—probably the most widely grown lipstick vine; it has dark green oval-shaped glossy leaves and orange flowers with some reddish brown markings.

GENERAL HEALTH CONDITIONS

Lipstick vines can be difficult to cultivate unless there is an adequate movement of air around them. Sometimes radical surgery is needed to remove some stems to let air enter the center of the plant. Plants tend to have many, many leaves, creating a mass of foliage where mealybugs and other insects can hide. *Aeschynanthus* varieties seem to bloom readily, even in somewhat shady places. A good pruning is in order once a year to make plants grow well. Never mist the lipstick vines.

GESNERIADS

Aeschynanthus lobbianus

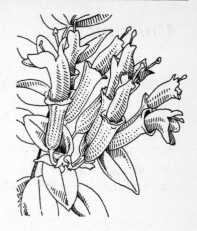

Aeschynanthus longiflorus

Aeschynanthus speciosus

Columnea

Columneas form a fine group of beautiful trailers or upright growers that may eventually reach 3 feet. The decorative foliage varies in size, color, and form: Some plants have buttonlike leaves, others 1- to 6-inch oval-shaped leaves. Tubular flowers are borne in leaf axils and are orange, red, or yellow in color. Recent advances in hybridization and cloning have produced some truly stellar columneas.

C. arguta—tiny dainty pointed leaves and salmon-red flowers.

C. 'Early Bird'—a trailer with somewhat large leaves and orange-red flowers.

C. microphylla—small leaves and red and yellow flowers.

C. 'Yellow Dragon'—displays large leaves and brilliant yellow flowers.

GENERAL HEALTH CONDITIONS

The hybrids are far more resistant to insects than the species, such as *C. arguta* and *C. microphylla*. These plants are plagued by mealybugs, so be forewarned. Spider mites also have a fondness for the little-leaved varieties; see part two for remedies.

GESNERIADS

Columnea arguta

Columnea 'Early Bird'

Columnea microphylla

Columnea 'Yellow Dragon'

Episcia

Only recently gaining favor with the public, episcias are furry-leaved plants with colorful, tubular-shaped flowers. The leaves in many varieties are highly decorative, with colorful markings. These plants like moisture and warmth and shading from the sun.

E. cupreata—the most popular episcia; it has fiery red flowers and copper leaves that are coated with white hairs.

E. 'Cygnat' has delicately fringed pale yellow flowers and furry leaves. *E.* 'Emerald Queen' is a vibrant plant, with emerald-green leaves and red flowers.

E. dianthiflora—white flowers and velvety green foliage.

E. lilacina—copper leaves and lavender flowers.

GENERAL HEALTH CONDITIONS

The beauty of episcias is not easily nurtured: Plants are terribly susceptible to mealybugs if growing conditions are too dry. Other insects that can attack the plants include spider mites; see part two for appropriate remedies.

GESNERIADS

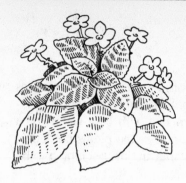

Episcia cupreata

Episcia 'Cygnat'

Episcia dianthiflora

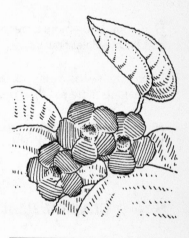

Episcia lilacina

Hypocyrta

Hypocyrtas are trailing plants with lovely shiny green leaves and pouchlike orange flowers that resemble miniature goldfish. After plants bloom, let them rest with little water for a month or so; then cut back the stems when plants get leggy. Watch for new varieties—these plants are just beginning to be hybridized.

H. 'Emile'—dark green shiny leaves and very bright orange flowers.

H. nummularia—more a creeper than a trailer, with tiny leaves and yellow-violet flowers.

H. strigillosa—has a very spreading semierect habit, with flowers that are more red than orange.

H. wettstienii—nice shiny green leaves and yellowish orange flowers.

GENERAL HEALTH CONDITIONS

Goldfish plants are sometimes attacked by mealybugs and scale. Thinning out helps eliminate problems such as spider mites. If you keep the plants evenly moist and in a shady place, they will most likely prosper.

GESNERIADS

Hypocyrta 'Emile'

Hypocyrta nummularia

Hypocyrta strigillosa

Hypocyrta wettstienii

Saintpaulia

The fine African violets, which need little introduction, have been enjoyed for decades. From the few African species of the gesneriad genus S. *ionantha* have come thousands of superior cultivars. Leaves vary in size and shape but are usually scalloped and somewhat hairy. Flowers are available in an array of colors, from flush pink to dynamite blue. There are countless hybrids; here are only a few of them:

S. 'Alkazam'—double lavender flowers.

S. 'Blue Fandango'—light blue flowers.

S. 'Flash'—double rose-pink flowers.

S. 'White Perfection'— large double white flowers.

Miniature

S. 'Honeyette'—red and lavender bicolors.

S. 'Pink Rook'—double pink flowers.

S. 'Tinkle"—double lavender flowers.

GENERAL HEALTH CONDITIONS

African violets, sorry to say, do have their problems: spider mites, which can wither leaves; and mealybugs, which visit often because the plants' leaves are so close together, preventing almost any ventilation. Water plants carefully with tepid water and try not to get moisture on the leaves, or they will stain. Keep plants in partial sun.

GESNERIADS

Saintpaulia 'Alkazam'

Saintpaulia 'Blue Fandango'

Saintpaulia 'Flash'

Saintpaulia 'White Perfection'

Sinningia

Better known as gloxinias, these handsome indoor-outdoor plants have slipper-shaped flowers in bright colors. The plants provide nice spot decoration. There are hundreds of varieties, most blooming at intervals throughout the year. After the plants flower each time, rest the plants for about 6 weeks, giving them little water, and then repot the tuber in fresh soil for another year of color.

S. barbata—bluish green leaves and white flowers that are streaked red.

S. concinna—a miniature with lavender flowers that are white-throated.

S. speciosa—the most popular species, with somewhat large leaves and pink or blue slipper-shaped flowers.

S. speciosa 'Cinderella' has red flowers.

S. pusilla 'Dollbaby,' a miniature, has lavender flowers.

S. speciosa 'Miss America' (not pictured) has double red flowers.

GENERAL HEALTH CONDITIONS

Gloxinias need some pampering; the soil cannot be too wet or too dry. Watch for mealybugs on the backs of leaves. Keep plants in a partially shady spot.

Sinningia barbata

Sinningia concinna

Sinningia 'Dollbaby'

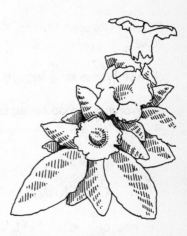

Sinningia 'Cinderella'

Streptocarpus

Streptocarpus is not a primrose, as its common name, Cape primrose, implies; it takes the name from its primrose-type leaves. These overlooked gesneriads are superb plants, with rich green foliage and funnel-shaped flowers. I can highly recommend some of the newer varieties—mostly basket growers— and the miniatures are also capturing hearts.

S. rexii—the most popular species, with rosettes of spatula-type dark green leaves and nodding white or pink flowers on short stems.

S. rexii 'Constant Nymph,'—a miniature with purplish blue flowers.

S. rexii 'Wiesmoor'—large frilled flowers (not pictured).

S. saxorum—a good basket plant, with dark green leaves and lavender flowers.

GENERAL HEALTH CONDITIONS

Do not be fooled by false claims—these plants do get infestations of insects now and then. Watch for mealybugs and red spider; if you see any, take the appropriate precautions as discussed in part two. Keep plants evenly moist, and do not overfeed them when they are resting in winter.

GESNERIADS

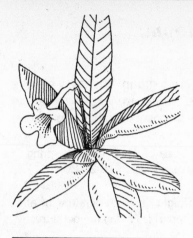

Streptocarpus rexii

Streptocarpus 'Constant Nymph'

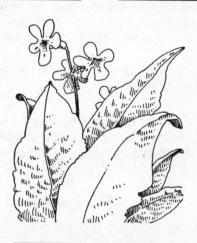

Streptocarpus saxorum

If Plants Won't Bloom:

The gesneriads are a diversified group and as we have seen with many, many plants, some are easier to get to bloom than others—but none is impossible to get into flower.

- *Aesychnanthus:* Give them buckets of water in early spring and increase feeding to twice a week using 10-20-20.

- *Columnea:* Provide a drop in night time temperature of at least 10 degrees (15 degrees is better) from day temperatures.

- *Episcia:* A very, very tough group to get to bloom; for some people they don't stop, but for others they refuse to bloom no matter how you treat them. As I have been only marginally successful with them, I plead ignorance here.

- *Hypocyrta:* Follow the same treatment as for *Aeschynanthus,* and plants usually bloom.

- *Saintpaulia:* A regular feeding program is the key to success with African violets; selecting the proper varieties helps, too. Ask your violet specialist where you can buy easy-to-bloom types.

- *Sinningia:* Try a little extra feeding after growth is well under way.

- *Streptocarpus:* If the plants won't bloom and they have a full head of leaves, wait a few weeks or a month. Then, if there is no sign of flowers, cut back . . . yes, cut back the plants to about the base and start again. Generally they come around the second time and bloom.

Orchids

LIGHT: Partial
WATERING: Keep soil evenly moist
FEEDING: Variable

In the last twenty years the epiphytic orchids have become popular houseplants, and rightly so. Orchids are beautiful, their flowers last a long time, and the plants themselves are generally amenable to cultivation, and are usually easier to grow than philodendrons. Orchids also make excellent cut flowers for the home. Most orchids have succulent-type leathery leaves. Most people think of orchids as purely tropical, but in the main they are temperate zone plants that enjoy average home temperatures and partial light (no direct sun). Grow orchids in fir bark. The best varieties are the following: *Cattleya* (florist orchid), *Cymbidium* (outdoor), *Miltonia* (pansy orchid), *Oncidium* (dancing-lady orchid), *Paphiopedilum* (lady's slipper orchid), and *Phalaenopsis* (moth orchid).

Cattleya

Years ago cattleyas were represented by the large and gaudy purple or white corsages. But no more; today most cattleyas are grown for the home, have smaller flowers, and come in every color except black. There is a wealth of beauty in this family.

In the *Cattleya* hybrids, the flower form remains the same, but colors vary, from a lovely white in autumn to a vibrant red at Christmas to a magnificent yellow in spring to an exquisite green in early summer and to orange in late summer! The selection is vast, varied, and beautiful. Ask your plant dealer for specific varieties.

C. bowringiana—early-to-grow; has clusters of flowers with fifteen to twenty flowers per cluster.

C. loddgesii—good and robust, with waxy pinkish white flowers; it is recommended for beginners.

C. skinneri—one of the old favorites; has clusters of small pink flowers in the summer and grows to 30 inches.

GENERAL HEALTH CONDITIONS

Cattleyas are rarely bothered by insects because the succulent-type leaves are just too tough for both sucking and chewing pests. Occasionally mealybugs may hide in the base axils, and if the atmosphere is too dry, spider mites may attack. Thanks to resistant varieties, disease is practically unheard of in these fine orchids.

ORCHIDS

Cattleya bowringiana

Cattleya loddgesii

Cattleya skinneri

Cymbidium

This versatile pot plant can be grown outdoors in favorable climates and indoors where winters are cold. The plants have been hybridized so extensively that there are miniatures (growing up to 2 feet), semiminiatures (up to 3 feet), and standards (up to 6 feet). There is a wide range of flower colors, from white to pink to red to green, with shades in between. There are thousands of cymbidiums, so choice is almost irrelevant—pick the color you like.

Cymbidiums need to be planted in a mixture of equal parts fir bark and peat moss; most nurseries sell cymbidium mix. Most plants require a cool period in autumn, to initiate buds (45°F at night for about 5 weeks), but some of the smaller, recently introduced cymbidiums, such as 'Pixie', do not need as much coolness as others.

Generally these plants like all the good things of life: good light, even moisture, ample feeding with a 20-20-10 solution in the spring and summer, and a 10-20-20 formula the rest of the year. Feed at 2-week intervals. Most cymbidiums adapt well to home temperatures and bear incredibly beautiful flowers that last for almost 2 months.

Miniature (to 2 feet)

C. Show Girl 'Marion Miller'—white flowers.

C. Ivy Fung 'Demke'—red flowers.

C. 'Oriental Legend'—reddish flowers.

C. Peter Pan 'Greensleeves'—green flowers.

ORCHIDS

Cymbidium 'Marion Miller'

Cymbidium Ivy Fung 'Demke'

Standard (to 6 feet)

C. 'Tapestry Red Glory'—red flowers (not pictured).

C. Voodoo 'Gypsy Red'—red flowers (not pictured).

C. Lilian Stewart 'White Satin'—creamy white flowers (not pictured).

C. Tiger Tail 'Canary'—yellow flowers (not pictured).

GENERAL HEALTH CONDITIONS

Cymbidiums can have their share of problems if not given proper care. Mealybugs might attack the grassy foliage, and slugs and snails are constant threats when plants start to bud. Also, a general cooling period to initiate buds is necessary (a 10 to 15 degrees drop in nighttime temperature) in early spring, or else plants bear little bloom.

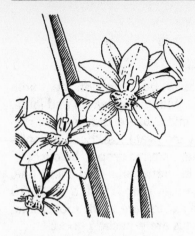

Cymbidium 'Oriental Legend'

Cymbidium 'Peter Pan'

Miltonia

These beautiful orchids are fussy, but I mention them here because they are so handsome. With grassy leaves and large, open-faced, pansylike flowers, miltonias need a growing medium of equal parts ground bark and peat moss. Keep miltonias evenly moist, and do not let temperatures drop to below 60°F at night. Do not place plants in drafts, and feed them only occasionally (once every month year-round) with a 10-20-10 solution. Hybrids have more vigor and are more robust than the species. The following hybrids are especially good:

M. Hanover 'Red Bird'—red flowers.

M. Bert Field 'Crimson Glow'—stunning red blooms.

M. Goodhope Bay 'Raindrop'—reddish pink flowers streaked white.

M. Bremen 'Anne'—white-margined red flowers.

GENERAL HEALTH CONDITIONS

Miltonias, of all the orchids, perhaps have the most susceptibility to insects and diseases. The grassy leaves are easy to penetrate, so they attract aphids and scale. And if the crowns of the plants are kept too moist, various types of rot can occur. Keep some charcoal at the base of the plants. Also, if the plants are not getting enough nutrients, the buds tend to cling to the stems and do not fully open.

ORCHIDS

Miltonia Hanover 'Red Bird'

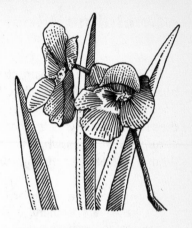

Miltonia Bert Field
'Crimson Glory'

Miltonia Goodhope
Bay 'Raindrop'

Miltonia Bremen 'Anne'

Oncidium

Oncidiums are the forgotten relatives of the orchid family, but why I don't know—they have beautiful small yellow and brown flowers that last for weeks. Although they are not dramatic, the flowers are certainly pretty, and most plants do very well in somewhat warm conditions (75°F during the day). Try a few—they will surprise you with their beauty.

O. ampliatum—a tough orchid that withstands much abuse; the leaves are leathery, and the brown and yellow flowers grow to one inch.

O. spathalaceum—flattened pseudobulbs, tall stem, many small brown flowers splotched brown.

O. wentworthianum—leathery leaves, thick growth, and masses of reddish-brown and yellow flowers.

Note: There are dozens of hybrids in the Oncidium group; most are relatives of *O. spathalaceum* and *O. wentworthianum*.

GENERAL HEALTH CONDITIONS

Oncidiums that are grown in good light and in a bouyant atmosphere should not be stricken with disease or plagued by pests. Once in a while, mealybugs may attack flowers, but usually the plants are fairly pest-free.

ORCHIDS

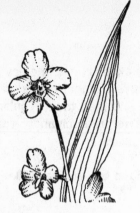

Oncidium ampliatum

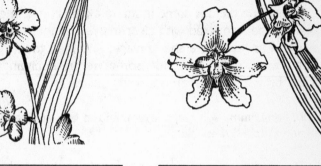

Oncidium spathalaceum

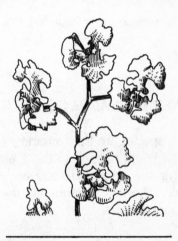

Oncidium wentworthianum

Paphiopedilum

These fine orchids come in many exotic colors; the flowers seem to be lacquered with clear varnish. Plant in equal parts soil and fir bark, in a shady place, and feed them regularly. There are dozens of hybrids; some are predominantly green, some brown, some maroon, some yellow.

P. fairieanum—soft green leaves, and in late summer, yellow and purple flowers bloom.

P. maudiae—a very handsome popular species with green lines on the white flowers.

P. hybrid—many varieties, form rounded, flowers in mahogany shades.

GENERAL HEALTH CONDITIONS

Paphiopedilum species are truly amazing: I cannot think of any problems that might happen to these plants. They seem immune to pest invasions and disease and grow well even under adverse conditions. Consequently they have become very popular indoor subjects.

ORCHIDS

Paphiopedilum fairieanum

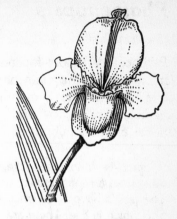

Paphiopedilum maudiae

Paphiopedilum hybrid

Phalaenopsis

Perhaps used more for indoor decoration than any other plant we know, the moth orchid is a stellar performer, with flowers lasting 5 to 6 months! And for all its beauty, it is incredibly easy to grow and rarely bothered by insects. Most varieties are now hybrids of *P. amabilis*, the original species.

P. amabilis—long, spatulalike, dark green leathery leaves radiating from a central crown, with white and very large flowers. The plants have a tendency to raise themselves out of their pot, which is fine. There are dozens of varieties, one prettier than the next.

P. 'Candy Stripe'—popular variety; white flowers, lined pink.

P. schilleriana—pink flowers; used in the hybridizing process to produce many different shades of pink.

GENERAL HEALTH CONDITIONS

These beautiful plants rarely expire because of insects, but if water gets in the crown, bacterial infections can occur and quickly kill the plant. And of course very cool temperatures may also cause foliage to die back.

ORCHIDS

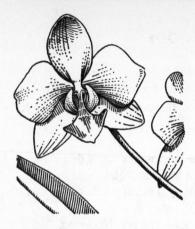

Phalaenopsis amabilis

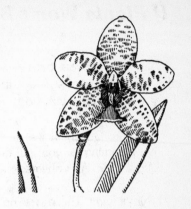

Phalaenopsis schilleriana

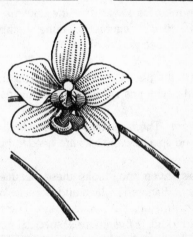

Phalaenopsis 'Candy Stripe'

If Plants Won't Bloom

What good is an orchid without flowers? Not much, so here are many ways to get these temperamental beauties to share their beauty with you.

- *Cattleya:* Queen of the orchids and rightly so; if these are not blooming for you, put them in a cool night location, say about 55°F. An unheated room or porch is just fine.

- *Cymbidium:* These are notorious for not blooming, but given a regular feeding program with special orchid food (from suppliers), flowers do come. And, yes, the nighttime temperature must be lower than the daytime temperatures by about 10 to 15 degrees. Ice-cube therapy sometimes works well, too (see part two).

- *Oncidium:* Let the plants crowd the pot; I mean *really* crowd the pot—and oncidiums will bloom their heads off.

- *Paphiopedilum:* These generally bloom with little trouble indoors, so no special gimmicks are needed here.

- *Phalaenopsis:* Everyone thinks these beauties need heat because they originated in Java and Borneo. Well, that may once have been true, but decades ago. The new varieties, crossed with other *Phalaenopsis* species, have produced plants that like it cool at night—about 60°F. In such situations they do bloom. Feeding *Phalaenopsis* too often can thwart bud formation unless you are very careful about when and how you apply food.

Palms

LIGHT: Partial
WATERING: Moderate
FEEDING: Moderate

As indoor foliage plants, palms are tough to beat for both beauty and ease of culture (most palms are almost impervious to insects). And there are so many *good* ones. Palms add a note of tropical ambience to any situation, and most have large and graceful fronds of pale green. Plant growth is fountain-shaped. Some palms reach heights of 8 feet or more. Generally palms do fine in partial light; water them heavily and then let them dry out. Feed palms only during the growing season—from spring to the following late fall—and not at all during the rest of the year. The best are: Areca (*Chrysalidocarpus*) (parlor palm), *Caryota* (fishtail palm), *Chamaedorea* (bamboo palm), *Howea* (Kentia palm), *Livistona Phoenix* and *Rhapis* (lady palm).

See page 132 for Areca palm.

Caryota

A popular and dramatic accent indoors, the "fishtail" palm has wedge-shaped leaves. These plants grow on a central trunk and branch out when they are about 3 to 4 feet tall. Do not panic if the tips of the fronds turn brown; this occasionally occurs in mature plants.

C. mitis—the popular one of the group; it is very handsome, grows tall, and has large shiny dark green fronds.

C. plumosa—sometimes called the dwarf fishtail, this palm is a handsome plant, although not as graceful as *C. mitis* (not pictured).

GENERAL HEALTH CONDITIONS

It is said you cannot kill a mature fishtail palm—and you know what? It's true. The plants are rarely bothered by any insects and even in adverse conditions have the ability to repel any disease.

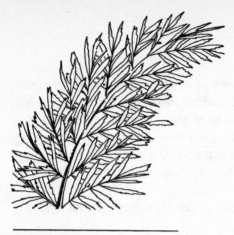

Caryota plumosa

Caryota mitis

Chamaedorea

Do not let anyone fool you about this group of plants: They are *the* best indoor palms. With bamboolike stems and graceful fronds, they can take abuse and *thrive*. Ideally they require good light and ample water. Feed plants moderately in the warm months, but not at all the rest of the year. Guard against mealybugs, which accumulate under sheaths on stems.

C. elegans—a fast grower, with dark green fronds and a somewhat compact habit.

C. erumpens—bamboo stalks and graceful dark green fronds.

GENERAL HEALTH CONDITIONS

These fine houseplants have one drawback: Because of the nature of the growth—sheaths over stems, as in bamboo—there is an ideal hiding place for mealybugs, and if conditions get too dry, the critters do move in. Spider mites are another attacker, you can sometimes see the webs. See part two for hints on routing these pests.

PALMS

Chamaedorea elegans

Chaemadorea erumpens

Howea

Howea is the expensive, glamorous palm everyone wants; it is fair to call it the queen of the palms. Fronds are shiny green leaflets on tall, arching stems. The plants need buckets of water during the warm months, but not as much the rest of the year. Do not confuse the true Kentia with the palm sometimes sold as Kentia: areca.

H. belmoreana—with its pointed fronds, this palm makes a very graceful appearance.

H. forsteriana—this palm, with very dark green fronds on beautifully arching stems, is robust and tough to kill.

GENERAL HEALTH CONDITIONS

These palms are very expensive, and if insect infestation occurs, or red spider or scale—and the plants are prone to these pests—you can lose a sizable investment. Be on the alert and catch trouble before it starts. (See part two.)

Howea belmoreana

Howea forsteriana

Livistona/Phoenix

Not often seen but certainly worthwhile in the indoor garden, both Livistona and Phoenix are attractive. Livistona has fan-shaped fronds, a heavy texture, and grows well in subdued light. Phoenix palms are almost impossible to kill and have sharp fronds, dense growth. Both plants rarely bothered by insects if kept in well ventilated area.

Livistona chinensis—handsome fan-shaped palms, leathery texture.

Phoenix roebelinii—somewhat large to 50 inches, sharp long fronds; grows dense.

GENERAL HEALTH CONDITIONS

I have grown both of these plants on and off throughout the years, and I recommend them highly, especially for gardeners who are somewhat prone to forget to water plants routinely. Flush with water a few times a year to keep fertilizer from burning leaf tips.

PALMS

Livistona chinensis

Phoenix roebelinii

Rhapis

Considered by palm devotees to be the best, *Rhapis*, also known as "lady palm," is truly beautiful, with segmented leaflets crowning tall stems. The leaves have a crepe-paper texture and are lined. Water freely in the summer but not as much the rest of the year, and keep in good light. *Rhapis* is rarely attacked by insects if the plants are grown in a well-ventilated spot.

R. excelsa— leathery, glossy green leaves with three to seven segments.

R. humilis—smaller than *R. excelsa* and more graceful looking.

GENERAL HEALTH CONDITIONS

I have a lady palm that is 20 years old, and other than leaf tips turning brown the plant has never been bothered by any type of insect or disease. Leaf-tip burn comes from too many nutrients in the soil, but flushing with water cures this one problem easily.

Rhapis excelsa

Rhapis humilis

Areca *(Chrysalidocarpus)*

This is one of the palms that frequently masquerades as the Kentia palm, but these stems are thin, and the fronds are less robust and graceful than those of the Kentia. The parlor palms needs plenty of water and then a good drying out. Keep plants only slightly moist throughout the winter. Watch for mealybugs on leaf stalks and at leaf axils.

> *A. lutescens*—somewhat shorter than most palms, this one growing to about 5 feet in height. Shiny green leaflets are borne on flexible stems (not pictured).

GENERAL HEALTH CONDITIONS

These plants, introduced into cultivation to compete with the more robust and famous Kentia palms, can be trouble-makers. They have a tendency to grow spindly if not fed enough and, as mentioned above, sometimes are attacked by mealybugs. Not my choice for a good house palm.

Other Plants

We have covered many plants, but some did not fit into our categories of plant families, so I'll discuss them here. These are the very popular plants that gardeners have grown for decades and that are still good. This group includes the following: *chlorophytum*, *ficus*, gardenia, hoya, and schefflera.

Chlorophytum

These grassy-leaved plants, popularly known as spider plants, grow in almost any situation and can be lovely specimens in large containers. Use the plants in hanging baskets for a dramatic accent. Plants like a poor soil (so add some sand) and a bright but not sunny location. Water the spider plant copiously all year; mild feeding once a month with 20-10-10 is fine. Don't fuss over these plants; they thrive on neglect.

C. elatum—the most popular species, with shiny green grass-like leaves and arching stems, with tiny white flowers.

C. bitchetii variegata—much the same as *C. elatum*, but the leaves are green striped with white, and the plant is somewhat more difficult to grow.

GENERAL HEALTH CONDITIONS

Chlorophytum is rarely troubled by insects; with ample water and light, it thrives. I have never ever seen a mealybug on my plants—and I have grown a few for over 10 years.

OTHER PLANTS

Chlorophytum elatum

Chlorophytum bitchetii variegata

Ficus

This is a large group of plants with diversified foliage and growth. Some, like the ever-present *F. elastica*, the rubber tree, have broad, leathery oval leaves. *F. pumila* is a creeping plant with tiny leaves, and *F. benjamina*, the popular banyan tree, has medium-size, oval papery leaves. Grow all species in bright light and keep soil consistently moist except in winter, when they can get along with less moisture. Do not overfeed plants; once-a-month feeding is fine. Use 20-10-10 solution.

F. benjamina—grows to 50 inches, with arching branches, glossy green leaves. The plant dies down in winter or early spring and loses its leaves. Don't force this winner into growth; wait for signs of new growth.

F. elastica—the favorite "rubber tree," with large, thick, textured foliage. Grows despite neglect.

F. lyrata—the well-known "fiddle-leaf fig", somewhat difficult to grow. Keep this one dry between waterings, and place it where there is an even temperature, since this plant hates drafts.

F. pumila (repens)—pretty plant with very small oval leaves, it is used extensively for topiaries. Easy to grow in shade, provided you always keep the soil evenly moist. Avoid feeding. Grows like a weed.

GENERAL HEALTH CONDITIONS

The *Ficus* genus offers so many plants that it is difficult to say that all are easy to grow or immune to insects. Generally the plants are not bothered by pests—the exception being the rubber tree (*F. elastica*); mealybugs do attack this plant.

OTHER PLANTS

Ficus benjamina

Ficus elastica

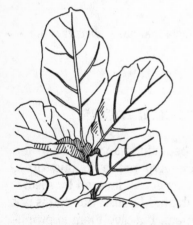

Ficus lyrata

Ficus pumila (repens)

Gardenia

Gardenia—one simply cannot leave out gardenias. They are tough plants to grow indoors, but people seem to like them so much that I am going to suggest some growing tips: Keep the soil somewhat dry; avoid fluctuating temperatures and drafts; do not move plants while they are in bud; and keep the humidity at about 50 percent. Feed plants with an acid food every 3 weeks—then hope for the best.

G. jasminoides—an old favorite with somewhat small flowers.

G. 'Mystery'—the most popular and easy-to-grow gardenia, with large flowers.

G. 'Tiny Tim'—a miniature gardenia that has flowers that are 1 inch across.

GENERAL HEALTH CONDITIONS

What can I say about these beautiful plants to soften the blow? Most of the varieties are very prone to bud drop, occasional bacterial rot, and are especially favored by mealybugs. I could say, "Don't grow them," I guess, but that is not really the answer. So cultivate them, but give them appropriate care and don't move them about when they are in bud. And put them in a location where temperatures will not fluctuate when they are budding. (See part two for preventive medicine.)

OTHER PLANTS

Gardenia jasminoides

Gardenia 'Mystery'

Gardenia 'Tiny Tim'

Hoya

These "wax plants" are old-time favorites, with leathery gray-green foliage and umbels of sweetly scented, waxy white flowers. Hoyas like an evenly moist soil, and only mature (2 year old) specimens bloom—the exceptions are the new minihoyas, which bloom when young. Give plants a bright location and repot them only when absolutely necessary. Feed moderately.

H. carnosa—leathery green leaves and umbels of lovely waxy white sweetly scented flowers.

H. motoskei—a climber (to 40 inches), with oval leaves that are spotted white; fine pink flowers.

GENERAL HEALTH CONDITIONS

Wax plants are not impervious to insects; mealybugs frequently attack them. Keep the alcohol cotton swab ready. They are rarely bothered by other critters.

OTHER PLANTS

Hoya carnosa

Hoya motoskei

Schefflera (Brassaia)

Scheffleras are tough tub plants that grow 5 feet in height, with large fronds of leaves. They need bright light, and you must allow the soil to dry out between waterings. Scheffleras do best in a somewhat shady location and should be fed with a 20-10-10 mixture during warmth months only.

S. actinophylla—large palmate leaves; can grow into treelike specimens. Keep the temperature a steady 68°F at night.

S. digitata—hairy, almost yellowish green leaves and needs a lower temperature than *S. actinophylla.*

GENERAL HEALTH CONDITIONS

These plants are seldom bothered by insects or disease and grow well with routine care.

OTHER PLANTS

Schefflera actinophylla

Schefflera digitata

Emergency Plant Protection Guide

How Plants Grow

If you know a little about plant structure and how plants grow, you will be better equipped to care for and protect your plants against insects and disease.

First, plants manufacture their own organic food from inorganic matter. All the organs of the plant—the roots, stems, leaves, and flowers—work together to create a living plant by storing the energy from the food.

Besides holding plants in the ground, the roots absorb both the water and the chemical substances necessary for growth. If water is not available to the roots, the plants die. Root tips are composed of cells that push forward from pressure in the rear of the roots. The root hairs on the tips force their way through the ground, seeking out moisture.

Behind the root tips is the growing area of the root; this is where the young cells multiply by dividing in two. As the plant develops, so do the roots: Big roots send out smaller ones that in turn develop into larger roots. The older roots carry water and nutrients to the plant and often act as water reservoirs, storing water for the plant in times of drought. The small, delicate hairs on the roots absorb water and food and are very sensitive to light and dryness; the hairs die quickly if they are exposed to air.

There is a one-to-one relationship between a plant's total leaf surface exposed to the sun and its total root surface in the ground. The food for the roots comes from the leaves. Thus, if the leaves suffer, the roots die back. Each leaf consists of three layers: the epidermis, the middle of the leaves, and the veins. The epidermis prevents leaves from drying out, the middle section contains chlorophyll (the plant's green pigment), and the veins contain elements that conduct water, inorganic salts, and foods.

The leaves make organic food out of sunlight, air, water, and earth salts. You'll remember the term from grammar school: photosynthesis—literally, "putting together with light." Via the process of photosynthesis, carbon dioxide, water, and the energy of the sun are changed into the carbohydrates and sugar that the plant can use in its life process. A leaf is therefore a brilliant chemical factory, combining gas and liquid. In one form or another, a leaf obtains oxygen and hydrogen (supplied by water), carbon (from the roots in the soil), liquid nitrogen, and other elements. The leaves then transpire water as vapor.

The stems store, distribute, and process the leaf and root products of almost all plants. Each cell in a stem must have a constant supply of moisture, air, and food. A stem stores and distributes the sugar for the plant and transforms the sugar into starch or other complicated proteins so that a plant can grow and live.

If all cultural elements are kept in balance—proper moisture, proper light, proper air in the soil, proper temperatures, and the correct foods—plants will grow and grow and grow. However, rarely can anyone correctly balance all the necessary ingredients, so sometimes plants do get sick—from lack of water, air, or nutrients, for example. And a sick plant is a plant prone to attack by insects. Included here is information on many different ways to help you help your plants, including knowing the enemy—the insects. There are many preventatives and remedies, enough ways to help you maintain healthy plants for years. Keep reading.

When leaves wilt, stems turn limp, or flowers fail to open, we immediately think that some insect or strange disease has targeted our poor plants for total destruction. In panic, we run to the nearest plant nursery and spend money for chemicals that are usually unnecessary and often do more harm than good to plants and people. (Whether you use these chemicals or not is your choice, I'll discuss them at the end of this book).

Dealing With Plant Troubles

Starting Right

Starting with good culture and catching trouble before it has a chance to gain a foothold are the best ways to maintain healthy plants. In their first days of residence on a plant, a few insects can be eradicated easily, but after a week or 10 days they may become a major force because insects breed quickly—very quickly. Mealybugs are capable of producing thousands of brood in a few days. So catching trouble means keen observation.

Pick off dead leaves and faded flowers—they promote disease—and keep plants well groomed by cleaning leaves with a damp cloth every week or so. This simple operation removes insect eggs and spider mites, helps promote shiny leaves, and keeps the pores of leaves open for air, sunshine, and proper humidity.

Plants That Do Poorly

Failing plants, with discolored leaves or premature dropping of buds, do not necessarily suffer from an attack of insects or disease. Poor culture can also harm plants. The plants may need repotting to facilitate good water drainage, a different method of watering, an increase or decrease in humidity, or simply a regular dose of fresh air. Check for the following symptoms and their causes before you reach for chemicals or perform any other preventative operations.

Symptoms	*Causes*
Leaves are not fresh green; weak growth	Too little light; too much heat
Brown or yellow leaves	Soil is either too dry or too wet; not enough or too much humidity.
Slow growth	Soil may not be draining properly, or plants may be in natural rest state
Leaf drop	Rapid changes in or extreme, fluctuating temperatures
Yellow or whitish rings on leaves	Watering with icy water
Bud drop	Plants shocked by a draft, or too-low humidity.
Dry, crumbling leaves	Heat too high; humidity too low
Collapse of plant	Extreme cold or heat

Insects

If all cultural conditions are good and you have purchased good, sound plants and they are *still* ailing, it is time to look for insects as the culprits. Here is how to recognize the bugs and the damage they cause, and also how to control them:

Symptoms	Cause	Control
Undersize leaves, spindly growth	Mealybugs; white cottony clusters in stems or leaf axils	Wash the plant with soapy water; rinse thoroughly with clear water. Or apply a cotton swab dipped into alcohol directly to insects.
Silvery streaks on leaves or deformed foliage	Almost invisible yellow or black-brown sucking insects called thrips	Wash plants with soapy water and rinse with clear water, or sprinkle paprika or pepper on soil.
Sooty mold on leaves or stems, stunted growth; sticky or shiny leaves	Oval-shaped insects called aphids; can be green, black, red or pink and commonly appear on new growth	Apply cotton swabs dipped in alcohol directly to aphids or give them a strong blast of water with a hose.
Leaves mottled or crumbly; shiny streaks on leaves	Almost - invisible spider mites	Spray plant vigorously with soapy water; rinse.

Symptoms	Cause	Control
Leaves turn pale, mottled	Tiny, oval, hard-shelled brown or black insects called scale	Try picking off the insects or scrub leaves with homemade nicotine solution (see p. xx).
Leaves stippled or very faded	Small flying insects called whitefly	Use a strong spray of water, or sprinkle pepper on the soil
Holes in leaves or edges of leaves eaten	Slugs and snails	Use beer or half-cut potatoes as bait

Diseases

Plant diseases rarely strike houseplants, but occasionally they do. Most diseases are not easy to get rid of, so be alert to the various symptoms so that you can prevent trouble before it happens. No one wants a costly plant ruined by botrytis or mildew—but a little knowledge can sometimes help save a stricken plant. Ailments that afflict plants manifest themselves in various ways: spots on leaves, rots at the crown, mildew on foliage, and so on. Many plant diseases cause similar external symptoms, so it is important to identify the specific disease in order to use the proper remedies.

As with insects, unfavorable growing conditions, such as too little or too much humidity or too much feeding, can contribute to plant disease. But generally diseases are caused by bacteria and fungus. Soft leaves, spots on foliage, and wilts are diseases caused by bacteria. Bacteria enter the plant through minute wounds and small openings and then multiply and start to break down plant tissue. Animals, insects, soil, and dust carry bacteria that can attack a plant. Also, in addition you can carry a disease from sick to healthy plants. Infected tools, such as scissors, can also carry disease so you're wise to sterilize tools with a flame or alcohol after using them on sick plants.

Like bacteria, fungi also enter a plant through a wound or a natural opening. Fungi spores are carried by wind, water, insects, people, and equipment. Because moisture is essential to their reproduction, fungi multiply very rapidly in damp, shady conditions. Fungi cause rusts, mildew, some leaf spot, and blights. Good air circulation goes a long way in keeping fungi and bacterial infection from striking plants.

Here are the main plant diseases and what to do when they infect a plant.

Symptoms	Cause	Control
Gray or watery green leaves, mushy crown	Bacterial blight	Remove infected plant parts and apply charcoal dust to the wounded areas
Coated white leaves	Mildew	Dust affected areas with charcoal dust
Gray mold on leaves, flowers	Botrytis blight	Discard infected plants
Leaves, flowers, and stalks spotted with concentric rings	Virus	Destroy infected plants, because there is no real cure

A detailed discussion of many safe preventatives follows to help you keep your plants healthy.

Alcohol

The regular rubbing alcohol you buy in drugstores is also a cure-all for almost any insect that attacks plants, including aphids, mealybugs, and scale. (Note: Alcohol *will not* combat ants, snails, slugs, or spider mites.) Alcohol is simple to use: Dip a cotton swab on a stick into the solution and apply directly to the insect. Since most of these critters are easy to see on plants, this is a simple way of eliminating the pests. Wipe away dead insects with a damp cloth.

I have used alcohol as a cure-all for insects on many, many different kinds of plants, from orchids to ferns. Here is my schedule for certain pests:

- Mealybug—apply three times a week for 3 weeks.

- Aphids—apply twice a week for 1 month.

- Scale—apply once a month for 3 months

To keep leaf surfaces—and vital leaf pores—clean, I use alcohol also as a wash. Wet a cloth—cheesecloth is fine—with some alcohol and gently wash the leaves. This makes the leaves glisten and keeps them free of dust and soot. After the alcohol rubdown, rub some clear water on the leaves. This simple beauty application does wonders for most plants, but do *not* use it on such hairy-leaved plants as hirsute begonias or gesneriads, such as African violets and *Streptocarpus*.

Aspirin

"Take two aspirins and go to bed" is the usual chant of the physician when you have a cold. And aspirin is also good for many plant ailments, such as invasions of whitefly, thrips, and springtails. Aspirin is cheap, does not pollute the environment, and even though it leaches into the soil, it does not harm plants in any way that I know of. Crush one regular aspirin into a powder and apply the powder to the surface of the plant's soil; water thoroughly. This will keep away the flying critters.

Sometimes plants, especially gesneriads, can suffer from too much feeding. The leaves become brittle, and the edges turn brown. As a remedy, flush the soil with an aspirin solution (two aspirins to a quart of water). Let the solution run through the plant soil several times.

Curatives From the Pantry

Ants are more of a plant problem than you may think. The ants themselves do not harm the plants, but they do "herd" other insects, using the insect's sugar to feed their own broods. Ants establish colonies of mealybugs on plants, and they occasionally herd aphids. Ants also spread disease.

It is not easy to get rid of ants in your home, greenhouse, or any area where you are growing plants. However, there are two good remedies: paprika and pepper.

PAPRIKA AND PEPPER

The ant repellents in small metal cases and other ant products generally work well, but they are expensive. I have had good success just pouring paprika in the places where the ants originate. When you see a string of ants, follow the line until you locate exactly where they are coming from. Sprinkle the paprika at these entrances; some component in the paprika evidently is not to the ants' liking, so they will leave the area. Pepper works on ants, too, but not as well as paprika. Ground chili powder can also be effective.

BEER

Beer really does eliminate most, but not all, snails and slugs. These critters eat small holes in leaves. Pour small portions of beer into bottle caps and place the caps in strategic places in the plant area. The snails and slugs will gorge themselves on the beverage and go away, there to expire from a grand hangover. I have had good luck with the beer method, though some people have told me it just did not work for them. But do try this method for your plants.

POTATOES

For years potatoes have been used by many gardeners to thwart snails and slugs. Cut potatoes in half and place the halves in areas where the culprits congregate. I do not know what it is in the potato that kills the slugs and snails, but this remedy does work. Or, try mixtures of water, sugar, and yeast.

COFFEE AND TEA

A few years ago I received several letters from gardeners asking me about the beneficial effects of administering used coffee grounds to ailing plants. Supposedly the coffee grounds contained ingredients that restored plants to health. Although I am a tea drinker, I did get some used coffee grounds and made a somewhat controlled test on a few plants. Three plants were given coffee grounds twice a week with their routine water supply and three other plants of the same plant families were given plain water, or a placebo, as doctors enjoy saying.

Ironically the three plants receiving coffee grounds fared worse than the ones with plain water; the grounds accumulated on the soil, making an unsightly mess, and when I did poke some holes to get them through the soil, the leaching of the grounds through the soil eventually clogged it. In the case of coffee grounds, the cure was worse than the ailment.

Having failed with coffee grounds, I went on to tea. In this case I used liquid tea, not the leaves, and once again set up a sort of controlled experiment. Three plants got liquid tea with their waterings and three plants (same varieties) did not. I chose plants that were somewhat ailing—yellowing leaves, limp growth, weak stems. After 3 weeks on the tea medication, I noticed three plants prospering—the three plants not receiving tea did not look as good as the other experimental plants. I continued this for several weeks; since I am a constant tea drinker, it was no bother for me.

After 2 months there was no doubt: The plants receiving tea regained health—stems were more solid and firm, leaves greener, and the entire plant took on a healthy look. I thought I had discovered a miracle potion, but not quite. After some research I discovered that what was occurring was that the tea liquid was an excellent leaching, or cleansing, agent that is to say, any plant receiving routine feeding (as my plants do) needs routine liquid leaching at 3- to 4-month intervals, and the tea did the job better than plain water. Thus I recommend tea therapy without reservation, whereas administration of used coffee grounds to plants seemed fruitless as a medication.

HERBAL REMEDIES

Many insecticides now on store shelves contain some herbs that repel insects. Pyrethrum is often used, as is ground tansy in different plant preventatives. These natural-based insecticides are usually expensive, and since tansy, pyrethrum, chamomile, and various mints are attractive plants in themselves, it makes more sense to use these in containers and put a few pots near or around your other plants. As a repellent against insects, these herbs work quite well. Indeed, they are certainly better than having noxious chemical odors in your home.

Of all of the herbals, I like tansy best because it is a handsome leafy plant and grows easily in a 6-inch pot. (It is available from many herb suppliers, ask for fern-leaf tansy.) Another good herbal repellent is rue. Few insects will congregate near this plant, and in a pot, rue, with its divided blue-gray foliage, is quite attractive.

There is no reason you should not plant up a few pots of rue and tansy—they do help keep insects at bay and thus ensure healthy plants. Give them routine three-times-a-week watering and feed once a month. Keep them in a location with filtered light.

Ice-Cube Therapy

I discovered the benefits of using ice cubes on houseplants many years ago. This type of treatment is a definite aid in bringing back ailing plants—those that always seem to be too wet or too dry. With ice cubes there is no hit-or-miss watering. If anything comes close to a miracle potion for plants, it's this type of slow watering.

Put five or six ice cubes on the top of the soil of a 5-inch pot; allow the cubes to melt naturally and forfeit routine watering. Use the ice-cube therapy about three times a week for small pots, more for larger containers of plants. Ice cubes generally melt slowly and the moderate-to-slow constant watering helps plants greatly to grow well. Soil never becomes really too dry or too wet.

When I go off on a few days' vacation, I always put some ice cubes on plant soil. This allows slow watering while I am gone and in general helps plants grow at maximum efficiency.

Soaps

Years ago, before poisonous chemicals were available, most people mixed laundry soap with water and used the solution to eliminate scale, mealybugs, aphids, and spider mites. This treatment is still effective, and inexpensive. And it keeps chemicals away from the scene. Use Fels Naptha or Ivory laundry soaps—*not* detergents, which can harm plants. Break off a 2-inch chunk of the laundry soap and add it to a quart of water. Mix up the solution thoroughly. Now either use a cotton swab and dab the solution directly on the insect or, if you are somewhat lazy (as I am), pour the solution directly on the affected area and then wash off the insects with clear water. Repeat the application about twice a week for 3 weeks to eliminate any eggs that hatch after your initial application.

Tobacco: Nicotine "Soup"

Tobacco causes much harm to humans, and it can also harm and even kill aphids, mealybugs, scale, and spider mites. Make a solution of used tobacco (get the tobacco from cigarette butts) and warm water, two cigarettes to a quart of water. Let the mixture steep for about 5 days and stir it before using it. Apply the tobacco mixture directly to the insects with a cotton swab; they will wither immediately. Keep up the treatment for at least 3 weeks, twice a week.

Resuscitation Chamber

This is a fancy name for a homemade enclosed plastic coverall, used to isolate an ailing plant and to provide the additional humidity that is generally necessary for restoring a sick plant to good health. (Signs of humidity that is too low are leaves dried at their edges or brown leaves.)

To make your own chamber, insert four 12-inch sticks into a sand-filled 12-inch square box, one stick at each corner. Now place the plant in the box and cover the sticks with a square of plastic film or a plastic bag. It is necessary for some air, but not too much, to reach the plant. Keep the box in subdued light.

When a plant is in a resuscitation chamber, observe it daily to see if it is reacting favorably to the additional humidity and the protection from fluctuating temperatures. Keep the temperature constant, about 75°F by day, slightly less at night (a slight temperature drop in the evening is normal). Do not spray or feed the plant while it is in the chamber.

Fracture Pins

I have had plants, especially bromeliads, suffer broken stems from falling off a bench or being accidentally struck. If this happens to your plant, do *not* amputate. It is far better to repair the fracture. For a minor fracture, insert small sections of toothpicks through the fractured area; or bind cotton string around the section to keep the stem or stalk in proper alignment. If the stem or stalk is severely broken, use both the toothpicks and the string, and apply tape as well.

It generally takes 2 to 4 weeks for the union to heal, at which time you can remove the splints, although I usually leave them in place. While the plant is healing, water and treat the plant as you usually would—no special considerations are necessary.

Leaf Amputation

Leaf amputation is removing injured, bruised, or cracked leaves. (In bromeliads, this condition is usually caused by temperatures below 58°F.) Do not just pull the leaves off, because doing so can injure the stems and rip cells. Instead, with a sterile razor blade (run a match flame under the blade to sterilize it), make a clean cut at the joint. Cut on the bias if possible so that the surface has more mass for healing itself. After you make the cut, burn a wooden match down to the nub or as far as you can hold it without burning your fingers. Let the match smoke out for a few seconds and then rub the blackened match end across the open wound of the stem. Cover the wound with an even layer of charcoal. (You can also use the packaged powdered charcoal sold at stores, but it is expensive, and wooden matches work just as well.)

Bypass Surgery

This treatment is the most difficult to administer, but it is not impossible—it just takes a little more time and observation. Usually, when a plant is not doing well (leaves shrivel and die) even though all cultural conditions are good, something is awry with the root system. Symptoms are leaves wilting, stems bending, and the entire plant sagging overall.

Uproot the plant by gently tapping the side of the container on the edge of the table to get to the root ball. Crumble away all the old soil and then spray down the roots with water to see which roots are white and which are brown. Clip away all brown and black roots. If the roots are not brown or black but the plant does exhibit the symptoms mentioned, crack any root at its tip. If the inside is green, the roots are still active. If no green is present, the root is dead and should be cut away.

Bypass root surgery involves a few hours of work. Once you cut away the dead or injured roots, spray the remaining roots with water and let the plant rest in a cool, dry place (about 70°F) for a day or so. Then repot the plant in fresh soil. Treat the repotted plants gently for a few weeks, giving it minimal watering and no direct sun until it shows signs of new growth and new vigor. At that time transfer the plant to brighter light, increase watering, and start to feed the plant lightly.

Artificial Light

Some years ago gardening under artificial light was very popular; begonias, African violets, orchids, and other plants inhabited many a cellar and closet under a barrage of fluorescent lamps. And where light was obscured in many apartments, artificial lamps were used. Incandescent lamps and fluorescent tubes do furnish the blue, red, and far-red light waves that are necessary for plant growth. So, if you have some plants that are not faring well in low-light situations, do consider using some type of lamps. Blue light enables plants to manufacture carbohydrates, red light controls their assimilation of these and also affects photoperiodism, a plant's response to the relative length of day and night. And far-red light is one of the colors that influence plant growth. Although we still do not understand all the ways light affects plants, we do know that the best sources of artificial light for plants are fluorescent and incandescent lamps.

Today there are a host of plant-growth lamps at supply stores. Some look like a standard light bulb and come with a bullet-type fixture as a package. The "growth" lamps are available under various trade names. You can also use fluorescent tubes for your ailing plants—these are made by several companies under different names. These lamps come in various lengths and wattages, and you need a canopy-type fixture for them. Artificial light units can be purchased, or you can make your own setup, provided that you are handy, using standard fluorescent lamps such as cool-white, warm-white, or daylight. Cool-white lamps worked the best for my plants when I was an apartment dweller.

Generally two 40-watt lamps 48 inches long are adequate to provide some additional light for plants that may not be growing as you want them to. Keep the top of the plants at least 6 to 8 inches from the lamps, and the lamps should be on from 12 to 14 hours a day.

If the fluorescent lamp setup is too large for the space you have, simply purchase the growth lights that look like standard light bulbs. These fixtures can be wall-hung and directed at plants, and they need only a small space.

Proper temperature is also a factor in success with artificial light; if possible, try to provide lower temperatures at night (by at least 10 degrees). Also, under lamps plants will generally need more watering and more food. Increase the watering schedule to three times a week and feed plants under lights every 2 weeks all year.

Most plants perk up considerably when put under artificial light, and a few months' sojourn may be all that is necessary to bring them back to normal health. It is good therapy for your plants.

Chemicals: If You Must

There are innumerable chemicals for protecting plants from insects. Just look at the shelves in your nursery—they contain a barrage of insecticides, miticides, and fungicides for use against specific insects or diseases. These chemicals come in granular, liquid, dust, or spray forms. Generally these toxic substances are rarely or never needed for indoor plants. Even the old-fashioned Nicotine 40, used as a cure-all on indoor plants for many years, is now on the banned list because it is highly poisonous. And too, the supposedly safe insecticides now available may prove to be toxic in future years. After all, it took many years before the old devil DDT was outlawed and many more years for chlordane to be removed from shelves.

Another problem with the use of chemicals is that the directions on the container can be ambiguous or confusing, often saying something like, "Use only on cloudy days," or "Use when temperatures are below 80°F," and so on.

If you must use chemicals, observe the following precautions. Always read everything on the labels and follow directions to the letter. Do not use beneficial insects, such as ladybugs or praying mantises, for plant protection because they will not stay put; they will end up in your neighbor's yard. Otherwise:

• Keep chemicals on high shelves, away from children.

• Keep all chemicals in one place, not scattered about.

• Never use a chemical on a plant that is dry.

• Never apply chemicals to a plant in a sunny location.

• Avoid oil sprays on orchids, bromeliads, and ferns.

• Isolate plants that are being treated.

• Apply insecticides outdoors if possible, and use a face mask.

Following is a list of the chemical "killers" so that you may be an informed consumer:

• *Chlorinated hydrocarbons* are deadly and include aldrin, dieldrin, heptachlor, and lindane. Most are banned, but some might still be available for sale.

• *Organic phosphates* include a large group of chemical killers, basically nerve poisons, such as malathion and diazinon. Both of these compounds are still considered safe by most authorities.

• *Carbamates* include the many fungicides and herbicides. These generally are broad-spectrum killers of mealybugs, cutworms, and so on.

- *Miticides* are highly toxic chemicals specifically used for spider mites.

- *Systemics* are the preventatives most commonly used because they control many insects. The material is applied to the soil and absorbed by the plant's roots through the sap; the sap is then poisonous to insects. Systemics offer good long-lasting protection, but they are very toxic.

- *Poison baits* are mainly for controlling snails and slugs. Most contain metaldehyde, which is highly poisonous. Sadly, I lost my dog to poison bait several years ago. Look for snail bait without metaldehyde.

- *Nematocides* are contact or fumigant poisons used in the soil to eliminate nematodes (worms) I think it makes more sense to use a *sterile* soil instead.

- *Botanical insecticides* are my choice *if* you have to use insecticides. These are ground powders made from the plants rotenone, pyrethrum, and ryania.

Afterword

It would appear, from the many books on plant care and from word of mouth, that plants have many, many ailments, but in reality problems can be minimal. Common sense preventatives such as those outlined in this book can keep plants healthy for years.

As this book goes to press, there are many dedicated people working to create plants that are disease-resistant and almost pest-free. And many new methods of biological control will someday provide plants with ample protection from pests and disease.

The main watchword is "A healthy plant is a happy plant and a happy plant is a healthy one." Keep growing!

A Word on Terms

Light, Watering, and Feeding Concerns

Light

There is great confusion about how much light a plant needs, and many gardening books compound this confusion with terms such as *filtered light, good light, moderate light, dappled sun, partial sun, partial shade,* and so on. So, for simplicity here is a simple equation:

"Bright" or "good" light = 3 hours of sun
"Partial bright" light = 2 hours of sun
"Partial light" or "Partial shade"; = Very little sunlight
"Shade": = No sunlight

The best sunlight for plants is in the morning because the sun is neither too strong nor too weak then. Late-afternoon sun is harmful to plants because then the light is very strong. Once again, moderation is the watchword.

Watering

"Water your plants when they are dry" is a safe dictum—but how do you know if they are dry? Simple. Test the top of the soil with your finger, punching down ½-inch into the soil to determine if it feels dry.

"Heavy watering": keeping the soil quite moist at all times

"Moderate watering": keeping the soil just evenly moist

"Light watering": keeping the soil just barely moist to the touch

Feeding

There really are only general rules for feeding plants, because plants such as aroids are heavy feeders whereas orchids require less feeding. Usually the best program is one of moderation—it is always best to give plants too little food rather than too much. For the sake of simplicity I can offer the following:

"Light feeding": provide food for the plant once a month all year except in the 3 winter months.

"Moderate feeding": apply food to plants once a week—very little during winter

"Heavy feeding": apply food to plants six times a month—very little during winter (perhaps once a month)

All plant foods consist of nitrogen, phosphorus, and potash. Nitrogen promotes good foliage growth; phosphorus produces healthy stems and growth and flower production; and potash aids in flower production. The contents are marked in this order on the package or bottle—namely 20-20-10 or 10-10-5 and so on. A moderate-type plant food is 10-10-5.

LEAF IDENTIFICATION

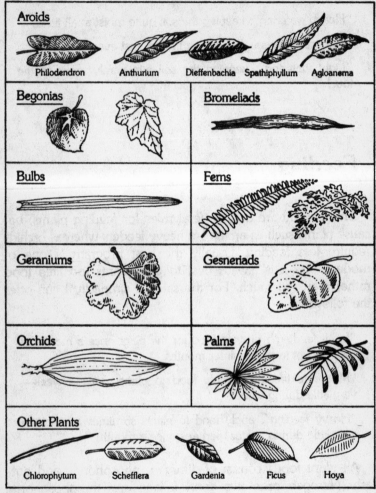

Aroids
Philodendron Anthurium Dieffenbachia Spathiphyllum Agloanema

Begonias

Bromeliads

Bulbs

Ferns

Geraniums

Gesneriads

Orchids

Palms

Other Plants
Chlorophytum Schefflera Gardenia Ficus Hoya

These are general identification leaf guides.
For more information, see specific plant drawings.
Cactii and succulents usually recognizable on sight.

PESTS

Spider Mite

Aphid

Mealy Bug

Thrip

Scale

White Fly

NOTES

Plant Name_____

Temperature _____

Water _____

Fertilize _____

Watch for_____

_____

_____

_____

Plant Name_____

Temperature _____

Water _____

Fertilize _____

Watch for_____

_____

_____

_____

NOTES

Plant Name_____

Temperature _____

Water _____

Fertilize _____

Watch for_____

_____

_____

_____

Plant Name_____

Temperature _____

Water _____

Fertilize _____

Watch for_____

_____

_____

_____

NOTES

Plant Name_____

Temperature_____

Water_____

Fertilize_____

Watch for_____

_____

_____

_____

Plant Name_____

Temperature_____

Water_____

Fertilize_____

Watch for_____

_____

_____

_____

NOTES

Plant Name_____

Temperature _____

Water _____

Fertilize _____

Watch for_____

_____

_____

_____

Plant Name_____

Temperature _____

Water _____

Fertilize _____

Watch for_____

_____

_____

_____

PLUME (0452)

HOME ON THE RANGE

☐ **FROM MY MOTHER'S KITCHEN: RECIPES & REMINISCENCES by Mimi Sheraton.** Food was the center of life in the Sheraton household and now Mimi Sheraton has compiled a book of her mother's wonderful delicacies that is full of love and spicy scents of good cooking. Over 280 kosher and nonkosher recipes are interspersed with a nostalgic journey back to childhood.

(256674—$7.95)

☐ **THE "SETTLEMENT" COOKBOOK 1903.** This facsimile edition of the 1903 *"Settlement" Cookbook* was originally subtitled "The Way to a Man's Heart" and it is still the way to any food lover's heart. This book will delight those interested in going "back to basics," those excited by authentic American recipes, and those who enjoy well-prepared food in any shape or form.

(256682—$6.95)

☐ **THE ORIGINAL BOSTON COOKING SCHOOL COOKBOOK by Fannie Meritt Farmer.** In 1896, Fannie Meritt Farmer, the principal of the Boston Cooking School, created a cookbook that, for the first time in history, provided carefully worked-out level measurements and easy-to-follow directions, leaving nothing to chance. This edition is a facsimile of the very rare first edition. "The Bible of the American kitchen."—*New York Times Book Review.* (261732—$8.95)

☐ **MISS MARY'S DOWN-HOME COOKING by Diana Dalsass.** Here are 120 authentic regional recipes from Lynchburg, Tennessee, lovingly passed down from mother to daughter and preserved by "Miss Mary" Bobo. For nearly 80 years this delicious fare has been a delight to anyone passing through Lynchburg. Some of the highlights include Hush Puppies, Fried Chicken, Raisin and Nut Loaf, and Fudge Pie. Also included are excerpts from Mary Bobo's personal story. (257301—$6.95)

Prices slightly higher in Canada

Buy them at your bookstore or use this convenient
coupon for ordering.

NEW AMERICAN LIBRARY
P.O. Box 999, Bergenfield, New Jersey 07621

Please send me the PLUME BOOKS I have checked above. I am enclosing $_____
(please add $1.50 to this order to cover postage and handling). Send check or money order—no cash or C.O.D.'s. Prices and numbers are subject to change without notice.

Name _____

Address _____

City _____ State _____ Zip Code _____

Allow 4-6 weeks for delivery.
This offer subject to withdrawal without notice.